GLOBAL MEGA-SCIENCE

To Adail

Best wishes

Dave Baker

GLOBAL MEGA-SCIENCE

UNIVERSITIES, RESEARCH COLLABORATIONS, AND KNOWLEDGE PRODUCTION

David P. Baker and Justin J.W. Powell

Stanford University Press
Stanford, California

Stanford University Press
Stanford, California

Printed in the United States of America on acid-free, archival-quality paper

Library of Congress Cataloging-in-Publication Data

Names: Baker, David, 1952 January 5– author. | Powell, Justin J.W., author.
Title: Global mega-science : universities, research collaborations, and knowledge production / David P. Baker and Justin J.W. Powell.
Description: Stanford, California : Stanford University Press, 2024. | Includes bibliographical references and index.
Identifiers: LCCN 2023038227 (print) | LCCN 2023038228 (ebook) | ISBN 9781503602052 (cloth) | ISBN 9781503637894 (paperback) | ISBN 9781503639102 (ebook)
Subjects: LCSH: Science—History. | Science—Social aspects—History. | Research—History. | Universities and colleges—History. | Education, Higher—Social aspects—History.
Classification: LCC Q125 .B26 2024 (print) | LCC Q125 (ebook) | DDC 509—dc23/eng/20231026
LC record available at https://lccn.loc.gov/2023038227
LC ebook record available at https://lccn.loc.gov/2023038228

Cover design: Martyn Schmoll
Cover art: Joel Filipe / Unsplash and iStock

To our families on both sides of the Atlantic

Contents

A Journey Across the Century of Science

This book was finished during the throes of the COVID-19 pandemic. As the world's media covered the terrifying emergence of a novel coronavirus and efforts to stop its global spread, another, originally less noticed, story has become part of the current zeitgeist. The spread of the pandemic was matched by the tremendous pace at which scientists generated new knowledge about the virus: its biology, contagiousness, therapeutic treatments, and the holy grail—effective vaccines.[1] Just six months into the pandemic, scientists had already published results from more than twenty-three thousand studies of the virus; this flow of new discovery had doubled every twenty days—certainly among the biggest explosions of research ever on a specialized scientific topic.[2] The devastation from the virus and mounting scientific knowledge about its origins and impacts are still ongoing.[3]

Obviously, the pandemic was highly motivating for the world's scientists, encompassing many disciplines and specialties, far beyond health. Yet this incredible scientific response did not come about by chance, and it could have been anticipated. Often missed in the coverage of this impressive response is the now massive and continuously expanding reach and interconnectedness of the world's scientific infrastructure, built over decades. This infrastructure serves as the backbone of an astonishing global capacity to undertake focused, frequently collaborative research at an unprecedented pace—now not only about a threatening disease but on literally thousands of topics.

Consequently, in 2020 alone, the world's scientists published over three million articles about their studies in more than nine thousand leading scientific and technical journals! These research articles, or "papers" in the everyday jargon of scientists, are where new discoveries, minute to monumental, are vetted and communicated. The sheer volume of papers reflects an extensive capacity for new discovery, increasingly done by globe-spanning networks of scientists carrying out research that continuously builds upon each other's findings. The pace of discovery and boundary-crossing worldwide collaboration are quintessential dimensions of what can be called *mega-science*. This exceptional change in the scope and dimensions of science—evolving for a long time, but not fully evident until recently—has inexorably transformed, for better or worse, the volume, breadth, and depth of knowledge production. The consequences of this vast growth and unparalleled collaboration for scientific discovery drive the centrality of science ever deeper into human societies in all regions of the world.

Yet mega-science was not supposed to happen, or at least not anywhere near to this level. As recently as the 1980s, prominent science-watchers predicted an end to the rising pace of science; it was, they thought then, completely unsustainable. Some pundits even went so far as to forecast a severe enough reduction in the world's capability for research to trigger a depression of the global economy. What occurred afterwards would have amazed them. Why their predictions were so wildly incorrect is because of a persistent misunderstanding of what facilitated the coming of global mega-science emerging at the turn of the twentieth century. Generalities about government policies and spending, geopolitical struggles, armaments, space races, economic demand, technological breakthroughs, and pressing societal crises are frequently trotted out as explanations.[4] And while these are certainly part of the story, none have the immediacy or consistent historical presence to be *the* foundation for what has transpired. They mostly play supporting roles, maybe necessary but not sufficient on their own to have set the far-reaching stage for global mega-science.

Instead, as argued and explored here, an often-ignored, yet major force behind science is the globally unfolding phenomenon known as the *education revolution* and its long-coming transformation of universities and their relationship to society, including science. This cultural process lies behind not only more people attending schooling for ever-longer phases of their lives than ever before but also, throughout the twentieth century and onwards, a steady inclusion of greater numbers of youth and young adults in universities and other postsecondary organizations across the world. In and of itself though, growing attendance alone could not have been the midwife to the mega-science we witness today. Also required was a concurrent influence of the culture of the education revolution on the very essence of what it means to be a university—and how this organizational form provides the crucial, well-resourced forum for the exchange of ideas and research for everyone devoted to scientific discovery.[5]

Over the same period, along with changing ideas about education in general, came the notion that universities could, and should, be places for generating new knowledge, science included. Of course, scholarship has always been part of the eight-hundred-plus-year history of the Western university, although mostly in a restrained fashion, with scientific research long considered beneath the lofty classical university of philosophy, theology, medicine, and law. At the middle of the nineteenth century something changed, first barely noticeable at a handful of universities, and thereafter, although contested and sporadic, spreading to universities across the world. Not only did universities become the organizational platform for teaching and research, but they also increasingly expressed their cultural power in recognizing and giving definition and boundaries to new areas of science, reinforced by discipline-based training and degrees in specific curricular areas. The cultural change born out of the education revolution and its new type of university was a melding together of the human energy and societal backing involved in expanding advanced education with a platform for the nurturing of scientific research. Of those approximately three million scientific journal articles in 2020,[6] the vast majority existed due to contributions of university-based scientists, frequently the sole source of even collaborative research efforts.[7] If earlier prognosticators of science had appreciated the maturing education revolution and its support of high-powered research occurring at most universities, they might not have missed the further explosion of mega-science that, even then, was brewing right under their noses.[8]

Usually, observations of a connection between education and science stop at noting that the former spreads scientific literacy and trains the talented few to become scientists. Required to sit through typical science courses in school, most people are more scientifically literate than were their grandparents or parents, and there are increasing opportunities to gain advanced scientific training. But this does not reflect the impact of the spread and growing intensity of science, trivializing what has occurred.

Alternatively, what we will refer to in shorthand as the *university-science model* formed a very potent and self-reinforcing connection between mass education and science, manifested in the symbiotic relationship between the university and science, without which it is unlikely that mega-science could ever have spread successfully to all continents and nearly every country.

Not long after the faulty forecasts of stagnant science, yet another tribe of science-watchers predicted that the university would lose its relevance to future science production. The notion was that many kinds of nonuniversity organizations, including science-based industries, would increasingly take over the knowledge-production enterprise, and thus universities would not keep pace in the science game. While the former turned out to be true, the latter did not. Even in an expanded environment with scientists working in many types of organizations, the university remains the powerful heart of mega-science, pumping out fresh ideas throughout society and providing the main platform upon which to research them. This is evidenced by the fact that, among a growing trend of studies from nonacademic organizations, such as businesses, most scientists in these organizations contribute to the research base by collaborating with university-based colleagues. Even the Internet itself was originally designed in universities to facilitate communication and collaboration among scientists across the world. And as will be shown, this university-based science has been cross-subsidized by the expanding role of advanced education within world society and the mass enrollments in postsecondary education in country after country that follow.

How Real Is Mega-Science?

Too often, reactions to the growth in volume of scientific papers dismiss it as merely reflecting hyperinflation in publication rather than real advances in discovery. Or some say that much of the rising volume of new discoveries could not have commensurate scientific value.[9] And there seems to be no shortage of new gloom-and-doom, end-of-science predictions, often based on scant empirical evidence. These knee-jerk reactions are untrue, especially considering that the incremental advance and development of fields over long periods of time is also essential to address current and future challenges, anticipated or not. Indeed, the present volume of papers directly corresponds to a concurrent explosion in the world's capacity to generate research. Employed scientists and researchers in all fields in wealthier countries tripled from 1980 to 2015 and grew worldwide by 50 percent over the past four decades.[10] While recently scientists have certainly been motivated to publish more, the fact that significantly more of them have been researching in more universities and other research-producing organizations over the century yields the steady increase in the rate of the world's output. Since the early twentieth century, for example, the number of graduate programs in science and technology and their volume of newly minted PhD scientists in the U.S. alone doubled many times. By 2015, about twenty-five thousand new PhDs in the sciences and related fields were graduating from the hundreds of research-active universities in the United States—every year. And in many countries, such as Denmark, Finland, and Israel, a once-miniscule research and development workforce now represents a substantial share of workers. With more opportunities as well as pressure for explicit and sustained collaboration among scientists, a multiplier effect of resources also leverages new scientific knowledge—and ensures it reaches wider audiences. Of course, with intensifying competitive pressure to publish comes some elevated incidents in errors, fraud, or misbehavior; for example, an estimated eighty-five hundred papers in 2005 among the over one million papers in STEM+ and social sciences journals were retracted, essentially a public notification that the paper is not of value, because of problems with the data, results, or other factors.[11] At the same time, reflective of the interconnectedness of mega-science is the growth in the sophistication in the identifying and tracking of such problems once found, and in replicating published results. Replication and synthesis of the huge volume of papers are perhaps even more important challenges to address; contemporary artificial intelligence (AI) transforms how we access vast quantities of existing knowledge, with dramatic technological and ethical implications.

As to quality, scientific discovery continually breeds novel areas for additional research, including, of course, high-value science arising from brand-new subtopics. This trend, part of the overall process of *scientization* also at the heart of the argument here,[12] is reflected in the creation since 1980 of some four thousand new major journals for papers on emerging subtopics of science, often at the intersection of disciplines, including parallel growth in journals publishing articles with the highest scientific impact. It is similarly reflected in the steady growth from nineteenth-century *fin-de-siècle* universities onwards in the creation of ever more hybrid academic departments and their graduate programs.[13] These continuously blend established topics of science into new research subfields to advance often multidisciplinary solutions to scientific and related social problems, and, of course, just to do more science. As with any robust institution within a specific cultural period, global mega-science becomes an intensified version of itself, an institution differentiated from the "inside out," or what can be referred to as the *scientization of science*.[14] This term reflects the greater institutionalization of a broadening scope of scientific inquiry through expanding domains and a deepening of the scientific activity within existing disciplines. In other words, although scientization is often taken to mean mostly the outward influence of scientific findings on nonscientific activities within society, the concept also reflects an emerging fundamental autonomy of internal organization and concentration of internal strategic actions by the thousands of contemporary scientists expanding this social institution globally as never before.

Such processes, farther down the line, also lead to robust applied uses that have revolutionized industrial production as well as human services. Imagine any area of life not affected by continuously advancing applications from basic research and its relationship to universities, as demonstrated by the smartphone or the Internet as a whole.[15] A concurrent spectacular growth has occurred in book publishing and patents, as the designs of useful, marketable things derived from the discoveries of basic research reported in innumerable papers. Universities and other research-producing organizations may capitalize on their discoveries, generating further scientization.[16] And all of these dimensions of mega-science are accompanied by vast—and increasing—global spending on research and development that grew at a rate faster than the world's economy over recent years, with larger investments, despite the pandemic's economic shock.[17]

Tracking a Cultural Model
Through Time and Space

The readily apparent size of mega-science fascinates, but the why and how of it remains in many ways a hardly anticipated story, one that when missed fosters profound misunderstandings about the world's capacity for scientific research—and its sustainability in the future. Before now, the cultural side of the education revolution has not been thought of as a foundation for mega-science. Histories of the university, of course, have chronicled some of the key transformations behind the argument here, such as the rise in the conferment of scientific and professional degrees, competition for faculty and its output, and the differentiation of fields as disciplines develop.[18] The growth of research activities at universities in selected countries is well known. Though useful, such trendspotting and historical reconstruction nevertheless lacks an appreciation for what caused the historical events in the first place. That requires tracking the development of the cultural ideas underneath the surface of the everyday understandings and motivations of the people and the expansive—and interrelated—social institutions of education and science. Like individuals, universities as complex organizations are also influenced by prevailing cultural ideas about what they should be and how they should operate. Increasingly rationalized in organization, their enactments nevertheless result from cultural ideas about goals and purposes, which at any one time form a socially constructed and accepted model that imperceptibly dictates organizational design and behavior.

Essential to understanding social change, cultural models also dynamically evolve. Their long-term consequences can be difficult to detect and interpret at any single moment or in any particular context. So we undertake a journey through a history of ideas behind the university-science model specifically and the scientization of society broadly across the long century of science from the late nineteenth century until the present, when university-based research paved the way for unparalleled advances in all scientific fields, not least communication and information technologies facilitating unparalleled collaboration or the vaccines protecting us against both common and rare or novel diseases. In various contexts, this model

7

became the main aspirational guide to how countries would grow their systems of higher education and reorganize their faculties for both teaching and scholarship, including increasingly specialized and pathbreaking scientific research.

Our journey starts in the drawing rooms of wealthy individuals and in the universities of Europe, particularly in the German-speaking region, moving west to the U.S. and then back again towards Europe and on to Asia and, ultimately, worldwide. We track a university-science model developed to such a degree over the twentieth century that most research-oriented universities are organized this way—everywhere from Berlin to Berkeley to Beijing.[19] Adopting some parts of the model at different points in time with various wrinkles from national history and culture, many thousands of universities and, recently, other postsecondary institutions joined the research game as well. Universities in countries that produced little to any globally accessible science before the 1980s, including those in Turkey, Brazil, Egypt, Qatar, Luxembourg, and Iran with their contrasting political systems and religious beliefs, now regularly contribute an appreciable flow of papers to the world's major science journals, usually in English. This widespread orientation towards producing research not only underpins the training of scientists but also makes universities the main place where research occurs. Faculty scientists are recruited not only nationally but increasingly internationally as universities become similarly guided by the universal research ethos. A distinctive shift has occurred that would have seemed odd even at the start of the nineteenth century, when most universities were primarily devoted to teaching and professional preparation, often centering on a traditional canon and serving state power. The global ubiquity of the contemporary university, with its similar structures and, most important, a similar culture based on the university-science model spread to all regions of the world, facilitates the intercultural exchanges and collaborations so essential to contemporary scientific advancement.

Following the pathways that the world's universities took collectively to reach mega-science, however, is not obvious. Assisting in the tracking of this expansive but underappreciated cultural model are guiding analyses of a unique and extensive set of information about the global flow of scientific papers from 1900 onwards. Analyzing the bibliometric information of the who, what, and where of 3.3 million published papers collected from the first year of each decade since 1900 offers a richly detailed map for the engrossing journey through time and space leading to what we call global mega-science. A more detailed note on data sources follows in the next section, including a brief description of the global collaborative process behind the project that well exemplifies the motivations, challenges, and benefits of transnational, intercultural, and multidisciplinary research.[20] In 1859, Darwin published his theory and empirical research in a book, *On the Origin of Species*; forty-one years later Einstein's first publication on relativity theory was in a paper among others in the journal *Annalen der Physik*; and since then the paper and the journal have been the main formal way scientists communicate about science.[21] At least since 1900, the volume of papers over time has arguably been the best direct metric of the flow of the newest scientific knowledge that is accessible worldwide. In their aggregate, the papers make up the sum of new findings on thousands of topics from basic science to advanced analyses of technology, engineering, mathematics, plus health, or what is now commonly termed STEM+. The 2015 wave of papers, around two million, was almost twice as large as the number published just five years before, and a volume that would have been unimaginable in 1900, when a grand total of only ninety-five hundred papers were published, largely by gentlemen scholars, in a small number of then-leading exclusive journals. Today, many universities and a few other large research organizations each produce more than that 1900 world volume of cutting-edge research every year. Since then, the annual volume of publication of scientific discoveries has increased by a staggering 21,000 percent, doubling its volume about every fifteen years.

Mega-science, though still not well understood, is very real, and it continues to spread everywhere and into further research fields. The flow of the millions of STEM+ papers analyzed is not a result of trivial change. Instead, it corresponds to a 120-year development in the ever-growing quantity of scientists across the globe, now collaborating in ways once unfathomable, thus expanding the awareness and often the quality of their increasingly jointly conceived and coauthored research. It has become routine to generate prodigious amounts of new discoveries and promising breakthroughs on a multiplying set of freshly scientized topics. And, as we will show, the ideas of the education revolution enacted through a changing university hatched and supported these trends with far-reaching implications for the nature of the world's science and its sustainability into the future.

Note on Data Sources and Collaboration

Funded by Qatar's equivalent of a National Science Foundation through a grant to Georgetown University at Education City in Doha, the assembled international team of the "Science Productivity, Higher Education, Research & Development, and the Knowledge Society" (SPHERE) project faced the formidable task of collecting enough valid and reliable information to reflect the entire development of mega-science, from 1900 into the twenty-first century, worldwide, and also information on the unfolding of the education revolution and university institutionalization. The team of sociologists of science and education, economists, and experts on higher education turned to several well-known scientific techniques, a full description of which can be found in this book's technical companion volume, *The Century of Science: The Global Triumph of the Research University*.[22] Briefly, first the team relied on the idea of a quantifiable indicator of scientific discovery, and others indicating advanced education development. Second, the team employed some sampling of discovery and years over the time period but endeavored to collect data for the broadest viewpoint possible. Last, some assumptions were made to make the task doable.

The team decided that scientific journal articles published in peer-reviewed journals is the most valid, historically consistent, and readily obtainable indicator of the volume of scientific inquiry at any one time, point, and place. Bibliometric analyses of science can include a range of additional published materials, although usually within a limited time and topical scope. Scientists do write books and government reports from time to time, and there are unpublished "grey" series of reports and correspondence that circulate online, but the globally recognized gold standard of declaring any and every discovery is to write an article on it and have it reviewed, vetted, accepted, and published in what is known as a scientific journal. The SPHERE project focused on published "papers," as they are routinely referred to by scientists, as a broad indicator capable of showing the transformation and global spread of the landscape of where science was produced.[23] There are other indicators—patents, R&D expenditures, and expert judgments—and of course no one indicator is perfect, since what each is ultimately asked to measure is very complex.[24] Especially in the contemporary era of increasingly coauthored papers, bibliometric counting is a conservative measure that reflects only a portion of myriad forms and results of research collaboration.[25] But papers offer a number of advantages over others, and important for the scope of our analysis, for the most part by 1900 published papers had become a highly recognized record of scientific discovery.

At the heart of the project, then, is an extensive dataset representing all *research* papers—omitting editorials, debates, conference reports, book reviews, and so forth—in STEM+ journals from selected years from 1900 to 2011. In fall 2012, the research team purchased the publication data of the Science Citation Index Expanded (SCIE) in an analyzable format from the bibliometric platform known as the Web of Science (WoS), maintained and marketed by Clarivate Analytics (formerly Thomson Reuters).[26] Data crucial for our analyses of the who, what, where of science—paper title, authors' institutional affiliations and addresses, journal citation impact factors (in later years), and subject area—every five years from 1900 to 1980 and every year from 1980 to 2012—were developed from the main dataset of papers. Since data files for 2012 were not finalized at the time of delivery, 2011 was the final year fully analyzed. Although the focus is on the time period of 1900 to 2011, we have recently augmented this with limited data up to 2020 and report these newer analyses when possible.[27] In addition to WoS, the Scopus database, maintained by Elsevier, has also evolved and expanded, along with the more inclusive (but less detailed and unedited) Google Scholar, and even "altmetrics" of nontraditional bibliometrics that complement traditional citation impact metrics.[28] At the time, WoS proved the team's best option and, in separate verification analyses completed since, very closely matches results from Scopus.

The purchased dataset sounded perfectly promising, but upon obtaining it the team quickly realized that the information prior to 1980 had limitations. Its completeness waned the further back in time the data went. Because the digitizing of journals proceeds in reverse chronological order, information from earlier years is less developed and often incomplete; many pre-1980 electronic files on papers did not include essential information such as affiliations or addresses of authors, and this usually was worse earlier in the century, when the scientific communities were small and providing organizational affiliation data was not standardized. For example, the proportion of papers without country of author information from 1900 to 1940 ranged from 56 percent to 90 percent. This is why contemporary analyses of scientific papers mostly focus on the decades since 1980 or even just the Internet era since the 1990s.[29] Another limitation was the breadth of STEM+ journals. Scientific journals come in all flavors—different languages, different subjects, different intended scope of readership, and different levels of selectivity of papers. Added to this is that no one knows for sure exactly how many journals there were worldwide at various points over the twentieth century, and while the ongoing digitizing and indexation by WoS, Scopus, and other platforms will no doubt become more inclusive in the future, historical analyses are particularly vulnerable to this limitation. Therefore, the team decided that it had to code the information by hand from the original hardcopy of the journal (or from microfiche) for pre-1980 journals to augment the WoS data, and because this represented millions of papers, it was only possible through a process of scientifically sampling a manageable amount and then weighting to estimate the total.

In many ways this is what WoS and Scopus have partially done anyway. WoS post-1980 is not technically a full census of all journals in the world. Instead, it includes a large volume of leading journals that have attracted more than 95 percent of the citations (cross-references) among scholarly articles. For example, in the most recent years our data is from all research papers in approximately ninety-two hundred such STEM+ journals, or the journals with many articles on discoveries frequently cited and thus built upon by other scientists (and with citations to articles in the same journal now a frequently used indicator of quality known as a "journal impact factor").[30] Some estimate that the WoS and other platforms are digitizing information about no more than a quarter of all academic journals on all topics, science and otherwise, but most agree that they are capturing the most influential, largest, most global, and most accessible (language-wise) journals across all STEM+ disciplines. Small journals pertaining to limited issues in one part of the world, maybe even just one nation, in a language other than one of the main ones of science publications tend to have miniscule readership and few citations, and hence are less likely to have had their papers indexed in these core bibliometric platforms. The main action of mega-science occurs mostly through the journals we sampled earlier in the century and that WoS had digitized post-1980. We try to remind readers of that in our subsequent analyses, but also note that the trends of mega-science are so clear that it is doubtful that inclusion of even more papers (in marginal journals; in many other languages) would much change the overall picture.

Connected to university libraries' journal archives around the world, the multinational team coded pre-1980 journal articles for the first and fifth year of each decade in a stratified, random sample of journals, and a random sample of research papers with each issue of a sampled journal. An initial sampling frame of all the journals for each year was used from the pre-1980 WoS data and stratified by four categories of science, technology, health, or all other. Also, in addition to English, for representation purposes the sample was stratified by German, French, and Russian journals, common languages for papers before English became the dominant publishing language of science.[31] The essential information was then coded by multilingual project staff, who if necessary used other archival sources to complete the crucial affiliation addresses of authors. This more comprehensive pre-1980 sample data, with weights, was merged with the WoS post-1980 data. A painstaking, multiyear process of countless staff hours yielded a uniquely comprehensive dataset of information on 3.3 million (weighted) of the world's scientific papers in major journals for a historical period that had rarely been represented by such data.

Obtaining indicators of the education revolution was easier. Gross-enrollment rates, the percentage of an age group attending a particular level of schooling, have been compiled for all countries across the full time period. More challenging though was that the actual number of universities has not been collected from each country over time (although we do estimate the numbers of those engaged in scientific research through author addresses).[32] To examine the interplay between not just higher education enrollment but also university creation, the team did exhaustive collections of foundings of universities in Germany, the U.S., South Korea, and Japan, and merged them with the historical papers dataset.

The team had to make some assumptions and rules for how to count papers and assign them to the university sector versus other organizational forms, and to countries and regions.[33] When papers have multiple authorships and cross-national collaborations—a huge trend driving mega-science particularly from 1980 onwards—this gives rise to significant not only technical issues in counting publications.[34] When counting total publications worldwide, the team used the number of unique research articles regardless of the organizational affiliation and address(es) of each article. That is, for global totals, single-authored and coauthored papers are counted once, regardless of the number of authors and countries involved. In other words, we do not double (or multiple) count collaborative publications for world totals. But when we compare across countries, papers are counted for each different country-based author on a "whole count" basis; for example, if a paper has two authors based in German universities and two based in South Korean ones, each country's total count would receive one paper. The same was done to estimate university involvement; papers with at least one university-based author were assigned to the university-sector count. Again, although some bibliometricians use fractional counting methods (in this example, each country receives one-half of a paper), we found that the large trends examined here are relatively insensitive to different approaches to paper counting.

Because of the significance of an author's country affiliation—not their actual citizenship status, but rather the host country of the research organization with which they are affiliated—for the analyses here the dissolution or unification of Germany, a major site of science and university development since 1900, required careful attention. Because of the lag time between research completion, article submission, and article publication date, a decision rule was adopted that allowed an article to be attributed to the former country up to three years after the date of transformative political regime change. For example, when occupied Germany was divided into the Federal Republic of Germany (West Germany) and the German Democratic Republic (East Germany), articles were thus coded.

During the period prior to 1949, all articles published by scientists in research organizations in the territories belonging to Germany were counted under "Germany." After reunification in 1990, articles from authors in both parts of the country were again attributed to "Germany," despite the different geographical borders. Similar rules were made for border changes among other countries, usually with contemporary borders applied retrospectively. Other coding rules related to historical and political changes in major science-producing countries, such as China's rapid rise to top producer of science after the end of the cultural revolution and the government's commitment to science and technology in 1978, can be found in the individual and comparative chapters of *The Century of Science: The Global Triumph of the Research University*.[35]

Last, over the later course of our project, more studies began to appear from a revival of the field known as "the science of science" using similar bibliometric data as we do here.[36] While to our knowledge none have examined the role of the university in global mega-science as we do, they report on recent trends such as growth, collaboration, impact, and so forth. Importantly, these findings corroborate the longer historical developments we found with the SPHERE project data.

CHAPTER 1

Professor Price's Error

The Rise of Global Mega-Science

Over three million new scientific journal articles now appear each year, evidence of the scale of scientization. Never before has the world been so capable of generating new scientific knowledge. Clearly, science grows and grows, but what the world has witnessed over the twentieth century and the first two decades of the twenty-first century is nothing short of a revolution. The consequences of this global scientization have far-reaching implications for humanity and the Earth. We are in the midst of a tidal wave of advanced scientific data, based on millions of experiments and studies, and forms of teamwork during each phase of research. From breakthroughs on particle physics, to the biological, chemical, and physical essentials of cells, to the engineering of things nano-small to enormous, to the uses of advanced mathematics leading to once unimaginable technology, to medical innovations of all kinds and everything in between, the pace and volume and organization of science has never been more robust. Along with unstilled curiosity, diverse challenges accelerated recent discoveries, but the pressure for solutions is not the only reason science continues to explode in volume. The coming of mega-science is more than yet another scientific revolution on a particular topic, it is nothing short of a steady transformation of the whole of science. It largely came about and was built upon a profound and symbiotic relationship between higher education and science that played out primarily within universities and between university-based scientists across the world. Only by understanding these relationships can we develop a clear vision of where science is heading into the future as well as its sustainability—with major implications for individuals, organizations, and societies.

Ironically, this exceptional, sustained increase in the ability to generate massive amounts of new scientific discovery of all kinds was once thought nearly impossible. Although early systematic estimates of new science revealed significant growth over the first half of the twentieth century, prominent science-watchers, starting around the mid-1960s, came to believe that such a rate of growth was simply not sustainable for much longer; perhaps only for another few decades or so. Surely, they hypothesized, a saturation would set in and growth rates would begin to decelerate. After all, the then world's leading science production rate of the U.S. was beginning to show signs of slower expansion; the former powerhouse of Western European science had sputtered during the World War II years; and the East Asian tsunami of scientific production was still decades in the future. So the thinking went that since science is a highly demanding undertaking requiring concerted expert effort, major financial investment, and substantial numbers of talented individuals working under the right conditions, it could not continue to grow. It made sense then to presume a coming leveling-off of science production—perhaps even stagnation. In fact, this "slow-growth prediction" became conventional wisdom and is still repeatedly predicted today about the near future. Such "wisdom," though, leaves us ill-prepared to understand how and why the world continued to increase its rate to the current unprecedented levels of new scientific knowledge, and how it is unlikely to significantly slow into the future.

Little Science, Big Science. Then What?

The well-known, yet faulty, prediction of a coming stagnation of scientific discovery can be credited to one brilliant man—Derek John de Solla Price. Unfortunately, Price's untimely death during a 1983 trip to London meant that he could never rectify the error in the central tenet of his otherwise enlightening life's work. A renowned professor at Yale, Price epitomized the world-traveling, elite scientist of the mid-twentieth century. Debonair with a shock of white hair and an ever-present, elegant hand-carved pipe, he began his career with a PhD in experimental physics. Upon discovering a passion for applying statistics to the accrual of scientific information, he earned a rare second PhD in the history of science from Cambridge. Price is credited with, in the mid-1960s, single-handedly inventing *scientometrics*, now a part of the study of bibliometrics within the booming field of information science. Defining it as "the science of scientific information," his research provides us with an initial systematic view of the rhythms of accumulating scientific discovery from the first millennium BC to the

1970s. And with novel analyses of then-esoteric records of discoveries from histories of scientists, scientific papers, and books, Price was the first to demonstrate that science could grow exceptionally fast. And it had consistently done so since at least the late sixteenth and seventeenth centuries.[1]

Republished three years after his death, calculations in his influential book *Little Science, Big Science* showed that by the 1960s indicators of scientific output were fully doubling their amount every fifteen to twenty years, a rate reflecting exponential growth.[2] Price likened his research to establishing a calculus of scientific growth, and with some hubris claimed that his results illuminated central "laws of scientific development." His first such law was that science grew exponentially. And without the data we now have, he assumed that growth from about 1900 to the 1930s marked a period of small-scale, mostly single-investigator discovery, which he referred to as "little science." Thereafter, over the middle decades, he assumed a shift to a period of larger projects, or "big science."[3] Then, since all exponential growth must evidently slow and level off at the upper limit of an "S"-shaped logistic curve, Price's second law forecasted that big science's exponential growth would halt in the coming decades. Not only was big science supposed to stagnate, Price predicted that a kind of inevitable scientific doomsday was lurking on the horizon. A world with too little scientific talent to follow through on new breakthroughs, based on the already growing pile of studies, would lead to too little new high-quality science and insufficient new discovery, all of which would supposedly be as harmful to the world's economy as another Great Depression!

Herein, though, lay his error. Even at the time of Price's death, world science was still experiencing exponential growth, and since 1980 there is no evidence of his predicted stagnation—quite the opposite. Our estimates of the number of worldwide journal papers from 1900 to 2015, as displayed in Figure 1.1, show historical growth as one long trend of expanding mega-science.[4] To give a better sense of the volume of papers over the last century, the first graph plots our estimates of the world's annual published science papers—growing from some 9,500 in 1900 to approximately 136,000 in 1960. The second graph shows the growth from the 1960s to the two million new papers published in 2015. Despite Price's

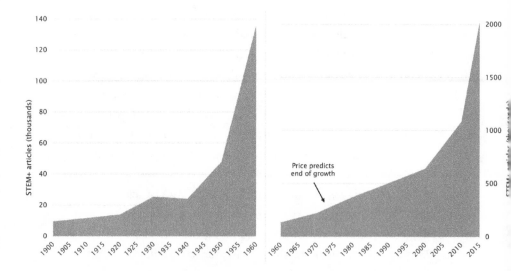

Figure 1.1. World Growth in New Scientific Knowledge, 1900–2015. Source: SPHERE database.

1975 predictions, the number of papers published annually continued to double approximately every fifteen years for the rest of the long century of science, just as he had shown it had done for centuries. Although the annual volume of papers earlier in the century was far less than after mid-century, there has certainly been no decline nor slowing; instead, the growth curve is exponential and continues up to the present day.

Nor is there evidence of any real breaks demarcating Price's assumptions about "little science" and "big science" periods, a categorization that our historical analyses will show is not necessary to characterize how science has developed. Instead of different periods, the qualities of mega-science were already in place by the earliest years of the twentieth century and mostly intensified in volume and reach thereafter. Of course, as the trends show, there have been periods of less rapid growth such as during the 1930s worldwide depression and World War II, but overall the record is one of sustained increases in the capacity to do more science by more scientists across more universities and other organizations in more countries.

To appreciate why Price's prediction proved so incorrect, consider an analogy to the biological concept of a "carrying capacity," or the level of resources in the environment that it takes to sustain a maximum population size of a particular species. Simply put, by assuming that the world's environment supporting science would not grow much after 1970, Price, along with other science-watchers of the time, underestimated the world's potential carrying capacity for science. A profound extension of the world's carrying capacity, however, was, and is still, playing out—largely, as will be shown, fueled by the resources and energy stemming from the education revolution along with networks of scientists working together to extend the cutting edge. Enamored with the fact that exponential growth of all types of things eventually ends, the originator of what would become the empirical study of science would not see an emerging global carrying capacity, supercharged by the expansion of the research university worldwide: a major source of support for research that was already fully evident since about 1900. The university has continued to provide the main platform for the growing amount of new scientific discovery and border-crossing collaboration and thus has postponed any end of growth for the foreseeable future.

WHAT IS GLOBAL MEGA-SCIENCE?

Reflected in the long-term growth rate in the volume of journal papers, at its core mega-science is *the rising worldwide capacity to produce more discoveries, occurring through a deepening connection between the university and science.* Four additional key characteristics of mega-science enhance this capacity, each also a result of the role of universities.

First, *a distinct trend towards greater inclusiveness and parity across world regions as to where new discoveries occur indicates the increasing globalization of science.* While the doing of modern science has always been somewhat open to all scientists and they have been mobile to a degree, never before have so many scientists in so many nations significantly engaged in research, autonomously and in collaboration, often far from their personal origins.[5] Plotted in Figure 1.2 is the shifting proportion of papers over the century by researchers working in different regions. In 1900, researchers in just 26 countries were producing all of the scientific papers in leading journals. Also, scientists in European countries, significantly led by Germany and followed by the U.K., France, and others, were producing about 75 percent of the world's papers, North American scientists were involved in just over 20 percent, and scientists in Asia and the rest of the world were publishing the small remainder.[6] From then on, two trends would unfold, resulting in ever-greater globalization. First, scientists in originally less-engaged countries became involved in new discoveries. It took until 1950 for scientists from an additional 9 countries to join the original 26 most productive ones, then over the ensuing three decades this exploded to 161 countries. By 2010 scientists from over 200 countries were regularly publishing papers in leading journals. It is precisely this growing world capacity that Price and others failed to anticipate. Second, shifts in shares of papers moved steadily towards greater parity. By 1930, European and North American scientists were authors on the vast majority of the world's papers. With its North American neighbor Canada, the U.S. dominated new discovery from the middle of the century on, but also with a steadily decreasing share as scientists joined in from other countries. Starting in the 1980s, the U.S. had average annual growth one full percentage point lower than the world average. On the other side of the world, by the 1980s, scientists in China, Japan, South Korea, Taiwan, and other East Asian countries had begun to contribute an increasingly significant share of the world's papers. This continued, such that among the 1.7 million papers appearing in leading journals in 2010, scientists in fifty European countries were involved in 35 percent of papers; those researching in the U.S. and Canada, by then the seventh most productive country in the world, were authoring about 25 percent; and another 25 percent came from fifteen Asian countries. The remaining 15 percent involved scientists working in relative newcomer countries to mega-science in the regions of South America, the Middle East, Oceania, and elsewhere. Also, as will be explored more fully in Chapter 9, it should be kept in mind that the later in the time period the analysis proceeds, the more likely papers have multiple authors researching in multiple countries across world regions.

Alongside gradually growing parity, a few countries still dominate the volume of paper output. In 2010, the U.S. remained the undisputed leader, with its scientists contributing to over a quarter of a million papers, followed by scientists in China, Germany, the U.K., and Japan (see

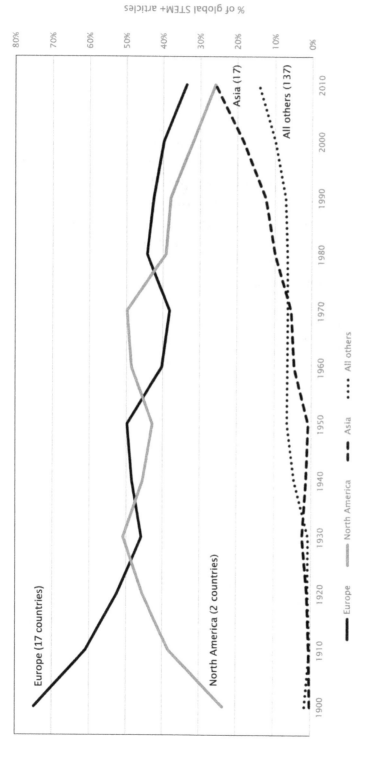

Figure 1.2. Increasing Globalization of Science: Regional Shares of the World's Journal Articles. Source: SPHERE database.

Table 1.1). But as discussed later, new 2019 estimates indicate that for the first time scientists in China authored slightly more than those in the U.S. Also, relative to the overall size of their populations, some countries produce far more than others. The range among the thirty most productive countries in STEM+ publications in 2010 is considerable: from 3.3 papers per 100,000 people living in India to 235 in Switzerland. Notably, the smallest, wealthiest countries are the most productive—all far above the mean of 91 papers per 100,000—led by Switzerland (235), Denmark (180), Sweden (177), Finland (156), Singapore (152), and the Netherlands (146).[7] These differences will be considered in assessing the sustainability of mega-science in the concluding chapter.

Second, *mega-science production is highly interconnected and thus increasingly transcends national borders.* As is commonly done, and as above, organizing publishing data by country or region blurs the true extent of the globalization of science. With the spread of the education revolution and the university-science model, mega-science does not really conform to the images of "American science" or "Chinese science" and is ever less a controllable, ownable property of any one nation. Of course, private R&D of applications of science can be, and are, treated this way and monetized, especially by firms, and countries' military and intelligence operations often attempt to control specific lines of research and scientists, usually without much effect on discoveries' evidential dissemination. For the most part, new discoveries are widely reported, now mostly through papers in digital form, to interested scientists across the world who integrate new findings in real time and adjust their own questions and tweak their experiments to achieve cutting-edge research. Try as they might, rarely can nations or companies hold the knowledge and exclusively use or sell it—even that science with strategic military value.

Collaboration is, more than ever before, a defining quality of science, as we find notable increases in scientists working together in ever-larger teams on research projects. This is partially what Price meant by big science, although contrary to his categorizations, collaboration has been growing since the beginning of the century of science. None of Price's contemporaries could have predicted how essential highly networked collaborations (mostly between university-based researchers) would become to discovery in most fields by the

Country	% of Total Papers	Papers per 100,000 People
U.S.	20.6	85.7
China	10.0	9.7
Germany	5.8	90.9
U.K.	5.4	111.4
Japan	5.0	50.8
France	4.2	85.6
Canada	3.4	128.9
Italy	3.3	71.7
India	3.0	3.3
Spain	2.9	80.1
South Korea	2.8	73.7
Australia	2.4	139.2
Brazil	2.2	14.6
Russia	1.9	17.3
Netherlands	1.9	146.2
Taiwan	1.7	93.3
Turkey	1.5	27.0
Switzerland	1.4	235.2
Sweden	1.3	176.8
Poland	1.3	43.4
Iran	1.3	22.0
Belgium	1.1	125.9
Denmark	0.8	179.9
Austria	0.7	115.0
Israel	0.7	123.7
Greece	0.7	76.9
Finland	0.6	156.5
Mexico	0.6	7.2
Portugal	0.6	73.0
Singapore	0.6	151.6
Mean (SD)	3.0 (3.9)	90.6 (59)

Table 1.1. Countries with Scientists Contributing the Thirty Largest Shares of the World's 2010 STEM+ Articles and Papers per Population. Source: SPHERE database.

end of the twentieth century. Coauthored papers by scientists from a single country are still the most common, but it is increasingly popular, a necessity even, for scientists from universities across countries to collaborate, and this is greatly facilitated by communication technologies. The sheer volume of such research collaborations is telling. By 2010, nearly all articles in the leading journals were coauthored. Of these, about one in four papers were produced through *international* collaboration, meaning that approximately a quarter of a million papers that year were authored by scientists working in at least two different countries.[8] Indeed, in numerous fields, collaboration is imperative and so ubiquitous as to be the standard process for most, if not all research in the near future.[9] This has become evident especially in meeting today's global challenges and the disciplines that address grave risks to humankind, such as sustainability science, artificial intelligence, and epidemiology.

Certainly, various levels of government, especially national governments, claim bigger roles in funding, policy, and at least symbolic propriety for their economies, but the qualities of mega-science make it difficult to capture discovery in national terms or to allocate rewards purely on national or even organizational bases. While, of course, scientists and funding agencies, even multilateral ones, still confront national barriers, geopolitical constraints do not much repress the growing pace of scientific collaboration. Teamwork is usually motivated and realized in everyday social interactions among scientists often committed to common questions and, increasingly, joint projects. Universities develop diverse strategies and programs to support their researchers to collaborate—to different degrees and in various forms of collaboration.[10] Regardless of past and current politics and conflicts, Iranian scientists work with Americans, Germans collaborate with Russians and Ukrainians, Chinese researchers with Japanese and Koreans, and every conceivable cross-national partnership now routinely expands contributions to science, raising visibility among diverse scientific communities. Obviously, American-trained scientists from other countries often do continue researching with their U.S.-based colleagues after returning home, but far beyond this expected cooperation is the phenomenon in which the U.S. and Europe, particularly their universities, have become international hubs of collaboration, with Asia's rise as a hub now a reality. Increasingly connected investigators from across the world join forces around common intellectual interests, skills, projects, infrastructures, research technology, and funding. For example, in 2020 the European Union announced a substantial €100 billion investment in research and development (2021–27) to be shared among researchers willing to collaborate across national borders to solve the most vexing scientific and societal problems. Many other EU programs also foster the extended carrying capacity, from international student exchanges to sponsorship of university networks in research and teaching, as they strengthen in-depth collaboration throughout educational and scientific careers.[11] But the massive growth of Europe's collaborative science reflects individual scientists' search for reputation and recognition as much as it does continuous EU promotion and funding.[12]

Given the increase in collaboration significantly enhanced by commonalities of the university-science model, as mentioned before the image of national science oversimplifies what has emerged as globe-spanning networks of researchers in all disciplines connecting around several regional hubs—and especially through growing numbers of research-intensive universities. As shown spectacularly in the global race to develop vaccines in the first year of the COVID-19 pandemic, the fundamental research carried out and shared by literally thousands of scientists everywhere provided the necessary platform for the concrete discovery pioneered largely by university-based researchers and produced by competing as well as collaborating pharmaceutical firms in different regions.

Third, *mega-science began its ascent from 1900 onwards.* As noted, our historical journey will show that the development of mega-science occurred over the long century of science. Instead of Price's distinct periods with supposed major qualitative differences, there was more quantitative intensification over time of the same trends. These culminated in mega-science spreading across much of the world and growing much faster in the last several decades than ever before, reflecting pure exponential growth.

Last, *with increases in discovery over the century, assisted by an evolving culture of science and education, mega-science creates the conditions for "scientization."* Science spreads and is applied to more dimensions through a multiplying of the number of topical areas of inquiry for an ever-expanding scientific infrastructure across universities. Reflecting this scientization, journals have increased the number of published articles per volume over time. Also, usually dedicated to new specific topics, STEM+ journals themselves doubled from approximately four thousand to over nine thousand in just three decades from 1980. By 2020, there were 9,526 of the most cited journals indexed in the selective Web of Science, ranging from *2D Materials* to *Zygote*.[13]

Mega-science is at the very center of the knowledge society, with its effects visible in diverse life-changing and life-saving technologies, from communications to medicine. While this may be obvious to many, there is surprisingly little systematic study of the various developments and factors that drive the entire enterprise of science—and therefore the foundation of the knowledge society. And while there are intense debates on the uses and misuses of science, we know little about the cultural forces leading this—increasingly global and networked—process in the first place.[14]

The Shifting Global Center of Gravity of Mega-Science

To plan the global path of our journey, the 120-year movement of the world's center of gravity of science reflects both the cultural spread of the university-science model and where and when successful innovations are being added. Using the information on where scientists work and their annual volumes of papers plotted against the planet's geographical space (oceans included) shows the cumulative pull of scientists in various countries and regions coming online to publish papers, usually while researching at universities and in other postsecondary research-producing organizations. The modern style of scientific publishing began in Northern Europe, and as displayed in Figure 1.3, by 1900 the science gravity point was situated west of Europe in the North Atlantic, just over a third of the way to North America. This point indicates the combined early dominance of the Northern hemisphere and the early German development of a new form of the university that intensified scientific training and discovery. With the initial emulation and innovation, over the next four decades the center moved steadily westward and slightly southward. This reflected the pull from the burgeoning science capacity of the U.S. and Canada, certainly supported by technological innovation and overall population and economic growth but equally driven by a somewhat unintentional adaptation of the European form of the university-science model into a wide swath of thriving public and private universities in North America.

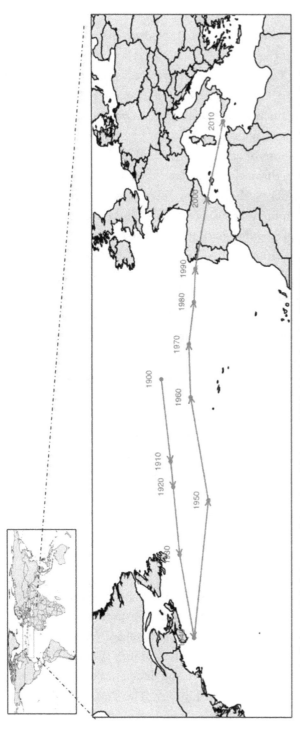

Figure 1.3. Shifting Global Center of Gravity of Science, 1900–2010. Source: SPHERE database. Note: Center of gravity is the geographical point, oceans included, representing the sum of the weighted geographic centroid of each country by total publications of that country in decennial years.

Then, starting after World War II the course of the world's scientific center of gravity turned around. For the next several decades, it traveled eastward. Rebuilt European universities and science systems powered this initial turn in the 1950s, and these small to mid-sized systems continue to significantly contribute to global science (see Figure 9.4). Starting in the 1970s, the course adjusted east by southeast, reflecting the pull from growing contributions of other countries and the expansions of their universities. Notably, this includes the later rise of China and other prolific East Asian countries, such as Japan, Taiwan, and South Korea, as well as those in the Southern hemisphere. No doubt at mid-century the easterly turn was imperceptible given the much publicized, highly celebrated dominance of American science in the post–World War II era. This led to an underestimation of the strength of Europe's diverse higher education and science systems that have proven their productivity and collaborative capacity, supported in recent decades by pan-European institutions and the continuous investment in European research and development. World media and governments continue to frequently embrace the image of the superpowers—the United States and China—pitted against each other in a race for science along with economic and military dominance. But this is far too narrow a picture. Instead, the great about-face of the center of scientific gravity and its rapid travel heralded the coming of a new world order of science—with multiple leading hubs of scientific collaborations and production powered by a global education revolution. As we will show, for example, universities in some European countries became as much of a collaborative hub for international authors of papers as American ones had been earlier. And, with an 18 percent average annual growth in papers from 1980 through 2010, China and its reformed universities have had a major impact on new discovery, just as the much smaller South Korea has done with its cutting-edge position in the education revolution and an even faster 20 percent average growth in papers. The concurrent new pull of publications from the Southern hemisphere, led by such countries as India, Australia, and Brazil, could be stronger, yet as will be described (in Chapter 8), the latter is experiencing difficulties in developing higher education to ensure its research capacity. The shifts emphasize the importance of key science-producing countries in the three major centers of Europe, North America, and East Asia as well as ascendant individual countries with rapidly developing higher education and science systems, explored in the following chapters.[15]

MAPPING THE JOURNEY

The intriguing 120-year path of the center of scientific gravity parallels the course of the education revolution, particularly regarding innovations in universities, and so charts our historical course in this book. Major stops in Germany, the U.S., Asia, and some countries that are newcomers to the process guide us to follow the university-science model's inception, development, spread, and scientific success. We chart the influence of major cultural ideas and their reshaping as they are translated across time and in diverse contexts. Each stop includes a story or two about individuals, universities or research institutes, and disciplines embodying the main historical and sociological development of mega-science at a specific place and time. Also incorporated are summaries of detailed analyses of the extensive science publication data, which fill in the bigger picture of mega-science's development embodied in each story. Since the general ideas of the university-science model were hardly fully embraced or made for a copacetic experience at every point, we also present examples of the conflicts and counter-ideas in the development of the university-science model. We study the wholly alternative models for science that could have been—and why they mostly did not take hold.

Before setting off, in the next chapter we revive earlier, prescient, but ultimately ignored ideas about the growing connection between education and science and its impact on postindustrial society. Then we begin with a brief visit to the "drawing rooms of science" of eighteenth- and nineteenth-century Northern Europe, the precursor to the university-science model. Next, traveling to a German university that was considered the world's best in the middle of the nineteenth century, we search for the gathering cultural ideas behind the university-science model and analyze how their implementation would become the prototype for the modern research university.

Moving westward across the Atlantic, we trace three stories to get there. The first is the American take on the university-science model, including a mix of mass undergraduate education, decentralized founding of universities, and flexible mission charters for scientific training and research. This process is encapsulated in a young American scientist's role in developing a proliferating form of a public university that reproduced many times over would become a significant engine behind that country's rising dominance of mega-science. A second tale, embodied in the rise of a smart student in early Maoist's China to become a Nobel Prize–winning physicist at an American university, shows how hundreds of thousands of foreign STEM+ graduate students like him along with thousands of talented American females and persons of color, previously ignored and even barred from entering graduate training, energized the research capacity of American universities as mega-science became ever more global—and inclusive. The last shows how the full breadth of American postsecondary institutions from prestigious wealthy private universities to huge public universities to, most recently, focused mainly on teaching, including four-year and even two-year colleges, now actively participate in research.

Then, with the center of gravity of science's about-face to the east, we revisit Germany's reestablished post–World War II universities that would by 1980 provide the platform for its scientists to achieve the world's third-highest share of new discoveries. There, serving as an informative counterfactual to our main argument, we explore what is known as the "dual-pillar crisis" of universities at the hands of an implemented policy of a limited, elite version of the university-science model. Named oddly enough after a Lutheran theologian, the rise of a culture of the individual "scientific genius" became the ideological foundation upon which were created prestigious, lavishly funded, nonuniversity, independent research institutes. Although this is an often-celebrated alternative to the university-science model, our analysis will nevertheless show that—if this model had spread worldwide as in Germany—it would have likely yielded only a reduced, unsustainable version of mega-science.

Moving east by southeast towards Asia—China, Japan, South Korea, and Taiwan—we follow the post–World War II forced decline of a centuries-old culture of selecting elites through advanced education that would be mostly replaced by the university-science model. Fueled by to date the world's most rapid adoption and implementation of the cultural ideas of the education revolution, significant university-based science production has exploded in this region over the past few decades. And contrary to the widely held opinion that these countries' success is because of unusual cultural practices, we show that their success is an intensification of what had been happening worldwide all along. Also instructive for understanding the causes of mega-science is this region's experience with developing "science excellence" universities, all of which ironically backfired but ultimately led to far more science production via a much broader array of competing universities than anticipated by the elite-focused policymakers.

To bring the global journey full circle, we make two last contemporary stops. The first is back to the U.S., where a physics professor at a large public research university collaboratively leads a massive astrophysics project of precise real-time research in Antarctica, conducted by three hundred astrophysicists, physicists, and engineers across fifty-two universities and other organizations around the world. Such sizable collaborations, along with many thousands of smaller-scale ones, culminate a multiplier effect of scientific resources that continues to extend the world's capacity to produce new science. And the university-science model has been at the heart of growing collaboration with first American, then European, and last Asian universities acting as major hubs for networks of scientists across the world to engage in elaborate research projects. Then, examining the implications of the contemporary era of networked science, we explore how collaborations across the globe continue to expand, pulling ever more scientists researching in ever more countries into mega-science. This is told via a final stop at recently created universities in two small countries, where two visionary and powerful women explicitly launched the university-science model to join the global scientific community. In doing so, they brought their countries into the globalized science of the twenty-first-century knowledge society. We conclude with predictions of the future sustainability of global mega-science.

Finally, we do not wish to make Professor Price the antihero of our story, as his work sparked the systematic empirical study of the historical development of science. So much conjecture about science, then and now, consists of cherry-picked examples of races to specific discoveries, juicy in-fighting among scientists, scandals of forged findings, and fascinations with Nobel prizes. Too often, sweeping conclusions about this massive and truly expansive enterprise are haphazardly made from such isolated micro-events or case studies of singular projects, disciplines, organizations, or countries. Thus we owe Price much for his pioneering work. Before routine computer computation with big data, he not only gave the world its first glimpse of the larger picture, he also introduced a new way to think about scientific development. Despite a mistaken assumption about future capacity, Price recognized the uniqueness of the century of science, so almost fifty years on his insights are still relevant. No one heralded the advent of these overarching trends until his analyses pointed the way, and the methods he developed, since refined by bibliometricians and information scientists, provide the foundations for our subsequent analysis. Nevertheless, without serviceable theory, Price and his contemporaries could not imagine the coming of global mega-science and the educational forces driving the stunning trends whirling up from their statistical calculations. We turn next to one man's prophetic but prematurely discarded theory about the role of universities in science and the knowledge society, which could have assisted Price and his fellow science watchers in anticipating the current dimensions of mega-science.

CHAPTER 2

Talcott's Prediction

Why a Century of Science?

As science grew across the twentieth century, many have speculated on the causes of that growth. Conventional wisdom, however, has yet to include the full answer. As noted before, factors such as government initiatives, economic interests, technological innovation, geopolitics, and even the utter weight of past scientific knowledge itself have all played their part in propelling new discovery to unforeseen heights.[1] Yet none of these thoroughly drove mega-science as have the education revolution and its spreading of the university-science model. As already noted, in the 1990s some science-watchers predicted that the university would lose its relevance to future science production. They reasoned that since many kinds of nonuniversity organizations, including science-based industries, would get into research, universities could not keep pace in the science game. Considerable hand-wringing by higher education specialists occurred over this popularized prediction, known as the coming of a "Mode 2" science—essentially an image of mega-science minus the role of the university.[2] And certainly it is true that due to the vast scale of research overall more science than before is now done outside universities. But assuming a declining influence of universities on mega-science was another miscalculation in predicting future trends: university-based research has not lost its share of the total. To the contrary, in fact, it has continually increased in importance. Indeed, predictions of an imminent decline of science production and less participation by university-based scientists simply have not occurred.

As a central indicator of this, Figure 2.1 shows that as the volume of STEM+ papers grew, so did the proportion of papers that included at least one university-based scientist as an author. Across the first two decades of the twentieth century only about a third of papers were from universities, as most were from independent laboratories, hospitals, firms, and even unaffiliated scientists.[3] By mid-century, the universities' share had grown to 50 percent and kept climbing, particularly rapidly in the 1980s, reaching about 85 percent of all papers by 2000, meaning that approximately 1.7 million papers included university-based scientists in 2015. And a lion's share of these were authored only by university-based scientists. Clearly not evidence of decline on the part of universities, this is the central consequence of the aforementioned unfolding of the university-science model that increasingly intertwined teaching with knowledge production in an expanding postsecondary education sector. Tracing the origins and implications of this way to organize universities is central to understanding mega-science and its interdependence with the education revolution.

THE SCHOOLED SOCIETY

The beginning of an answer to "why mega-science and its universities" starts with one man's novel and prescient prediction, developed nearly a half century ago. Over forty-six years professing social theory at Harvard University, Talcott Parsons developed a theory of human society in the grand European tradition. Balding, with a short-cropped mustache and kindly eyes, at the height of his influence Parsons looked like an amiable WASP-y uncle from a 1960s TV series rather than a world-leading social theorist. Arguably though, he was the most renowned and original of America's intellectuals in the then relatively young discipline of sociology—the scientific study of society. Trying out his theory to consider the causes of modern society, in the early 1970s near the end of his life Parsons made two observations and from them derived a unique prediction central to our tracing of the ideas behind the university-science model.

First, by this point in the U.S. and increasingly elsewhere, more and more children and youth attended school for longer. But instead of making the common observation that as society becomes more complex, more education is needed, Parsons turned this notion on its head. Instead, what he saw was the result of an unfolding of a revolutionary new cultural value that humans should receive unprecedented amounts of formal education. The idea to educate extensively most, and soon all, members of human society had become so strongly attractive that older rationales for education were no longer necessary. This was transformative in the sense that no former human collective had strived to provide formal schooling, in such cognitively demanding amounts, to nearly everyone. Until

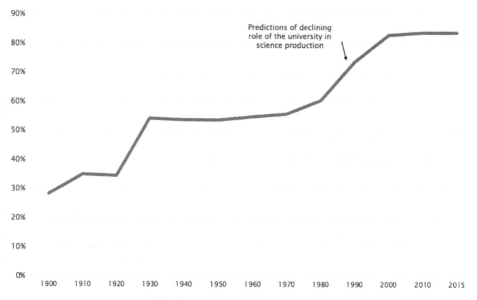

Figure 2.1. Increasing Role of University-Based Scientists in STEM+ Discovery.
Source: SPHERE database.

mass education began to take root in some societies in approximately
the nineteenth century, only the children of elites in most societies might
be formally educated, including those destined to lead religion and a few
other special small strata of individuals. Mass education via compulsory
schooling of nearly everyone had never been tried before. And, as Parsons
labeled it, an *education revolution* of this scope changes individuals and
society alike in heretofore unimagined ways.

Second, Parsons observed that at the cutting edge of this revolution
was a thriving higher education sector, and, most saliently, the emerging
twentieth century university heavily engaged in discovering new knowledge.
But, he observed, the university was not leading this sweeping change in
a top-down commanding fashion—nor was it because of some national
plan.[4] Rather, he saw in the take-off in enrollments in American colleges
and universities the creation of waves of new types of individuals, ideas,
and, most important, the capacity to create new knowledge, scientific and
otherwise. The effect of mass education in primary and secondary schooling was certainly creating youth ready for higher education. Yet Parsons in
addition saw that the growing acceptance and use of the university rein-

forced and legitimated the value and enactment of the education revolution itself, including its connection to knowledge generation. A widening and deepening education revolution had since at least early in the twentieth century set the foundation for the university's role in global mega-science.

While it would take time for economic and sociological research to fill in the full picture of what was happening, Parsons was clearly on to something unique. His sociological consideration of this came at the right moment—precisely in the middle of unprecedented increase in education attainment with strengthening norms about what made for a normal education, and growing income pay-offs from educational degrees in the American economy. The average total years of schooling attended by youth rose from just finishing primary school in 1950 to finishing secondary school and going beyond—only thirty years later (as shown in Figure 2.2). What would become a repeated pattern in many other countries over the rest of the century, the U.S. was early to expand secondary schooling and venture on to more postsecondary education for a significantly wider part of the society. By the late 1950s and 1960s, the country's large postwar baby-boom generation would experience an environment in which obtaining a high school degree was the norm, capping a forty-year effort in growing primary and secondary enrollments that would, at the end, sweep many youth into significantly more educational attainment than their parents. And this was nearly equally true for females and males, and increasingly, although not fully, for children of minoritized racial, ethnic, regional, and socioeconomic groups. Education became, gradually, increasingly inclusive for diverse ability groups, extending the human right to education for all and inclusive education.[5]

Then, this growing secondary education was met by a growing supply of capacity at two- and four-year postsecondary institutions. Consequently, while less than 10 percent of twenty-five- to twenty-nine-year-olds had earned the BA in 1950, this more than doubled to 25.4 percent of young adults by 1975. Also, the resulting increase in human capital among young American workers plus major technological innovations would create a secular trend of increasing wages (adjusted for inflation) for high school and BA degree holders up to about 1970, all along stimulating further education attainment within the population.[6] Although education con-

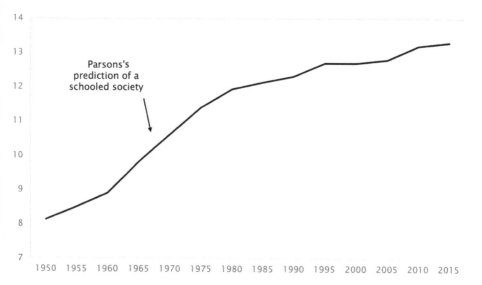

Figure 2.2. Rapidly Arriving Advanced Schooled Society: Average Years of Education Attained by All Adults in the U.S., 1950–2015. Source: Our World in Data, https://ourworldindata.org/grapher/mean-years-of-schooling-long-run.

tinued to expand post-1970s, particularly for the BA and post-BA, never had so many made such a qualitative jump in attainment relative to earlier generations.

On the basis of these two insightful observations, Parsons uniquely predicted that an unfolding education revolution, with its ever more culturally relevant university, would yield an intensified form of postindustrial society in the U.S. and eventually across the world. He noted four related processes underlying his sweeping prediction. First, that proportionally more individuals would find their way into higher education, and as a result colleges and universities would increasingly be seen as places to educate the whole individual. Although notions of former liberal education included educating the whole person at least for a select few, mass education incorporated an explicitly human development approach into a logic behind educating all. Second, that an already thriving research capacity in science, social science, and humanities at universities would greatly increase in the near future. Third, that the kind of knowledge most coveted would be more abstract, theoretical, and universal, qualities precisely aligned with the essence of science. Fourth, that such knowledge would become seen as legitimately applicable to all manner of problems facing human society, from the physical, economic, and environmental to the psychological and deeply intimate, a broader phenomenon that includes the notion of scientization. These four processes would continue to interact symbiotically to make the university a central institution extensively shaping advanced-modern society. And indeed, schooling, including the university, has become more cognitive in content, less narrowly vocational in intent, more focused on broad human development, and increasingly connected to universal ideas of knowledge.

A subtle, yet crucial, distinction that Parsons added was that universities' rising volume of new knowledge over the first half of the century—modeled ever more on the terms of scientific theory and empirical methods—was not necessarily accomplished in the service of practicality; rather, he saw it the other way around. The university's deepening cultural legitimation and hence success in academic research established and expanded and privileged the social utility of such knowledge. While this may seem obvious now, before Parsons was commissioned in 1972 to reflect on these observations and their consequences in his last book, *The American University*, coauthored with Gerald M. Platt, few had thought about education as a *revolution*, much less as a leading social institution of the coming postindustrial world. Nor had anyone looked deeply enough into the university's capacity to significantly alter the dynamics behind discovery that came to underlie global mega-science. At least not exactly like this. A few other leading intellectuals at the time, among them Daniel Bell, Burton R. Clark, and Neil Smelser, made related observations. And the mounting growing pains from enrolling America's post–World War II baby-boom generation in colleges and universities regularly appeared as popular media stories, but no one so accurately perceived the society-wide consequences of the intensifying university as did Talcott Parsons.[7]

Of course, a kind of cultural schizophrenia about the university parallels its increasing centrality—much celebration mixed with concern that mass higher education might weaken its educational value. Why this occurs, and interestingly appears exactly as the education revolution matures, are beyond the argument here and have been considered elsewhere.[8] Nevertheless, it is fair to say that postindustrial society is heavily invested in the belief that universities are highly legitimate places where authoritative knowledge (especially scientific knowledge) is generated and integrated with the creation of academic degrees and experts who then carry new understandings about topics into all aspects of everyday life. Whether this is always good or best for society is not the main question here. Instead, the point is that universities do not merely train people for occupational roles. They also have a major, autonomous role in creating new forms of society in what is privileged, how it is socially stratified, how jobs are done and profit made, and what becomes important and what recedes into the background, and in influencing many other aspects of social life. Interpreting educational trends this way anticipated a deepening interconnection among assumptions about what the university can, and should do, that continues until now. Powerful in its simplicity, this belief, increasingly taken for granted, strengthens connections among labor markets, economies, and educational credentialing, and, in turn, universities' growing mission of knowledge production. In growing relevance of the university, Parsons saw a deepening interdependency between the discovery of scientific knowledge and more activities within society, remaking the university as the ultimate cultural authority on knowledge.[9] An emerging greater centrality of the university in society pulled more people into higher education to meet proliferating goals. And as will be shown here, over the twentieth century this provided essential resources and energy for development of the university-science model and mega-science, earliest in Germany and parts of Europe, then in North America, and eventually in country after country as scientific research globalized.

Parsons's observations and predictions about the education revolution and the university, and ultimately the foundation for mega-science, came out of his general theory of human behavior, culture, and their interaction to form a viable society. It is a sweeping theory proposing to explain nothing less than the rise and development of modern society. Along with the theory, he also sought to establish an abstract form of theorizing as a kind of enduring system of sociological concepts for future scientific inquiry about society. Unfortunately, this made Parsons's texts notoriously inaccessible. Crafting overly complex prose in the odd belief that it would make the argument sound more scientific, his writings are so needlessly dense that demonstrating one's fluency in arcane "Parsonian" became a sport among sociologists of the time. When superficially read the concepts appear tedious and obvious, yet cutting through the compulsive abstraction and endless fourfold concept schemas reveals an impressive set of ideas about how human societies seek stability while often changing towards greater complexity. This is essentially a solution to a central sociological challenge of simultaneously explaining social integration (stability of culture) and social differentiation (change towards greater societal complexity).

Prior theories had tended to account for either societal stability or social change instead of tackling how both can exist at the same time and even escalate one another. Parsons's breakthrough insight was that both can, and often do, result from the same unifying process—the institutionalization of culture. Parsons reasoned, in what was a parsimonious hypothesis and a significant step forward, that while human societies must meet essential functions to exist and flourish, the anthropological record shows that they can accomplish this in strikingly different ways with varying, yet meaningful, social actions expected of members of societies at different historical periods. And these different paths to similar ends across societies originate out of a remarkably diverse array of ideas, beliefs, and values, or in short, the full breadth of known human culture.

Developing and employing various aspects from this array, societies are made up of clumps of culture known sociologically as social institutions, which are organized around functions such as production, protection, leadership, kinship, education, and knowledge creation.[10] The use of the terms "institution" and "institutionalization" here should not be confused with their conventional meanings of a specific brick-and-mortar place, as in the "University of Oxford is an institution," or institutionalization as a process of affiliation within such a brick-and-mortar place. Instead, social institutions are clusters of culture animated through individuals. They are conceptual mental mappings or schema defining ranges of how to think, feel, emote, and act around social functions. In telling the story here, "social institution" and "institutionalization" will always be used in this Parsonian sense. Social institutions provide the cultural models for what is normal and hence abnormal (that is, antisocial) behavior, emotions, and motivations, thus integrating all of the infinitely possible actions of individuals, their personalities, and repertoires of behavior and feelings within overarching functional patterns of a particular society. Regardless of the specific content of their institutions, when societies effectively accomplish a deep integration across them, they thrive, and if not, they fragment and collapse (meaning not so much annihilation but as abandonment of failed cultural forms). In other words, societies become more stable as their institutions—being justified ways and ideas about accomplishing important functions—are believed in enough to carry the weight of a commanding reality guiding what people actually do, think, and feel. At the height of the Roman Empire, for example, military conquest, genocide, and large-scale enslavement for material production and societal survival were much believed in for some central functional processes, and were celebrated and thus integrated across the empire's other institutions of religion, social stratification, leadership, family, and so forth.[11] Yet an institution of domination and production by these means is now mostly unacceptable—increasingly amoral and asocial—while contrastingly mercantilism, capitalism, and wage-labor are mostly taken-for-granted contemporary realities in many regions of the world.

Perfect integration across institutions is never possible, because in striving to achieve stability, societies do not generally remain static. They tend to change towards greater complexity over time—developing ever more elaborated areas of human activities and their shared meanings. This paradox of change through stability is what Parsons's theory seeks to explain. He argued that the inexorable gravitation towards greater social complexity is more a product of the same sustaining cultural process than anything else. Technological advance, disruption from other societies, environmental change, pandemics, and sporadic internal upheavals can, of course, durably change societies. But these are not the most common ways complexity increases. While these dramatic events of sudden change receive a lot of historical attention, across the long human record societies have mostly diversified through the essence of what makes them possible in the first place—the tendency to live by, and thus to intensify, cultural patterns guiding how to collectively function. Thus the more a way of doing important functions is taken for granted (that is, integrated), the more its logic takes on the power to explain more of reality (greater institutionalization), thus resulting in more complexity.

A theory of human society elegantly based on the single social process of institutionalization of culture stands in contrast to older, less sociologically satisfying hypotheses about the effects of natural resources, technological gain or loss, dynamics of production, or simplistic biological evolution. Contrary to persistent criticism of his ideas, Parsons did not assume that conflict was out of the picture of social change. Instead, he strove to go deeper—beneath symptoms of change to their core processes. Likening culture and its institutionalization to a kind of societal DNA, his view was a major step forward in advancing the central driving idea behind the scientific study of humans and their collectives, namely that the dynamics of human society, along with the reality of living within one, is socially constructed through culture maintaining functional social complexity and social solidarity.

To understand the rise of particular social institutions, and hence specific societies, one must identify those sets of cultural patterns that underlie and in fact drive their historical development. Parsons was fascinated with extending this approach to perhaps the most challenging of all incidences of cultural dominance and change—the historical transition to "modern society." A theoretically satisfying account of the worldwide spread and disruptive power of the major institutions of modern society has been, and remains, the scientific holy grail of sociology and related social sciences. First, approximately in the seventeenth century in northwest Europe and then elsewhere, a notably different society with new intensifications of defining social institutions became apparent. Called by various names such as the "rise of the West," Western society, liberal society, modern society, and industrial society, all refer to the same emergence of what has become a globally dominant form of culture with steady change towards unprecedented social complexity and challenges to solidarity now widespread enough to have pushed aside many other cultural forms (that is, premodern societies can no longer sufficiently integrate around their institutions in the face of modern culture).[12]

To explain the emergence of modern society, Parsons argued that human society was transformed as never before by three cultural revolutions. Being Parsons, "revolutions" of course did not imply the usual notion of events of violent political upheaval; rather he meant radically new patterns of cultural meanings, values, and ideas about doing essential social functions—new intensifying social institutions. Violent episodes certainly could be catalysts of social change, but again, most social change does not result from sudden disruptions. Instead, Parsons meant that new institutions, often over long periods of time, revolutionize prior cultural arrangements to meet central functions in new ways.

The first two—the capitalist-industrial revolution, an ever-intensifying institution of production and material distribution, and the democratic revolution, a similar intensifying institution around leadership and polities—are now well established as part of the cultural origin of modern liberal society. And Parsons recognized that a cultural revolution of mass education is just as central as the other two, thus forming a triad of institutions driving this change. By education revolution, Parsons means the spread of the cultural idea of more education for more individuals for more reasons, followed by the organized action of schooling, all added to the well-known tendency of modern institutions to culturally validate rationality as the "best way" to accomplish things. Slowly at first, this revolutionary idea spread and intensified the centrality of the university as a—indeed, *the*—main legitimate institution in knowledge production.

This, then, is why, at the end of his quest for a theoretically satisfying explanation for the coming complex liberal society, Parsons focused on education, and particularly on the Western form of the university as its most culturally advanced version. In short, his prediction foreshadowed what can be now called a "schooled society," in which a culture of formal education is a dominant and highly legitimating process pushing rationality, cognitive behavior, and, with them, scientific knowledge production into more aspects of life.[13] In fits and starts, certainly unevenly, and with resistance as is particularly evident now, much of the world moved towards what is often referred to as a "social democratic-capitalist society." What Talcott Parsons astutely observed and predicted a half century ago was that the full cultural pattern should be referred to as a "social democratic-capitalist-educated society," or even perhaps a "social democratic-capitalist-university" society. And his prediction has proved more accurate than even he could have imagined. Global mega-science, and indeed the whole reality of a knowledge society, arose from the cultural cradle of the education revolution and its growing complexity in the form of unprecedented science capacity, driven chiefly by the expanding number of the world's universities.

By his death in 1979, this renowned intellectual and recognized founder of American sociological thought was being supplanted by a generation of sociologists obsessed with Marxist theory of economic conflict as the sole driver of change. Parsons died the day after giving an address celebrating the fiftieth anniversary of his PhD from Germany's Ruprecht Karl University of Heidelberg (est. 1386; Germany's oldest). At his alma mater, he anticipated the rise in cultural significance of the education revolution by speculating on whether education-based classes (opportunities and rewards distributed by educational attainment) would replace in significance the most sacred of Marxian concepts—the economically based social class. Although a brilliantly insightful harbinger of what was to come, his synthesis of social integration and differentiation became nevertheless unfashionable. It was even prematurely dismissed to the point that generations of new sociologists are instructed in a basic Marxist critique of Parsons instead of learning much about his actual theory. And not surprisingly with such intellectual shunning, the study of his ideas about the revolutionary impact of education was also pushed aside, at least for a time.

This has led to the education revolution being somewhat adrift in the conversation about its role in postindustrial society. Images of educational expansion as overeducation, credentialism, and even a kind of collective rip-off reappear over and over, even in the face of overwhelming empirical evidence to the contrary. Evidence abounds about the role of education, beyond just credentials, as a main ingredient of human capital in transforming economic hierarchies of common economic outcomes, such as wage differences, that stratification and social mobility research employs to indicate adult status. The historical upshot of much economic research is that investments in educational development and rapid periods of increases in educational attainment in many countries yielded aggregate increases in human capital that have changed job contents, firms' strategies for profit-seeking, and the organization of the workplace. Analyses of aggregate economic output confirm that a significant dynamic behind the growth of the American economy over the nineteenth and early twentieth centuries was the nation's ability to expand primary and secondary educational opportunities, elevate the level of cognitive skills of workers, and match them with rising technical job content and profit-making in industries. Beyond human capital in the economy, similar amounts of evidence indicate that what is learned and skills gained through formal education influence all types of aspects of the individual, and in the aggregate are partially responsible for worldwide change in patterns of population health, demographic regimes, and the historical rise in cognitive capacities, to name several.[14]

It is no simple feat to accurately predict the future, and mostly what has happened since vindicates Parsons. Many of his ideas are currently enjoying a revival in social theory circles.[15] How else to consider the worldwide boom in education at its most advanced levels; the explosion of knowledge, particularly scientific; and, most important, the unexpected deep comingling of these two trends? As no social theorist had before him, he placed the steady growth in formal schooling for ever-wider swaths of populations and the development of the university at the center of the rise of a knowledge society and economy based on scientific discovery by an integrated, self-reinforcing process. In short, the education revolution launched the university into its role as the main organizational platform and provider of resources and legitimacy for the capacity to generate global mega-science; precisely an argument that would have helped Price and others make clearer sense of their observations of exploding rates of scientific discovery.

ALTERNATIVE ARGUMENTS

All good theorists push their argument exclusively, and Parsons was no exception. And our journey to mega-science and the rise of the university-science model will be a partial test of a major implication of his theory. At the same time, it would be naïve to assume that the education revolution and its universities were the exclusive facilitator of mega-science. Other large-scale factors, such as money, geopolitics, and technology, are clearly empirically associated with growing science. But these influences are only part of the story, and some are themselves rather distant from the factors close to the research process. And they are also themselves partial outcomes of growing scientization. If the world had only become materially richer, organized by nation-states, with greater technological production and armaments, it is not a given that mega-science would have flourished. In a sense, without the education revolution, the social revolutions of capitalism and representative democracies would not have yielded the full array of integrated social institutions of the twentieth century's intensification of modern society. Our intention is not to imply that these more studied factors are irrelevant to mega-science; instead we focus on the less-appreciated, but crucial university-science argument.

Also, there are other popular theories about what drives science, some of which are completely orthogonal to the education revolution and universities. While we will not test these, it is useful to mention and briefly critique them in light of the trends of mega-science shown in our analyses of papers. A common idea is that science is a collective good, and one that society "needs," particularly industrializing societies and now postindustrial (high-tech, complex, intricately differentiated) societies. And there are, of course, patents, reflecting applications of science, based on both old and new science, and evidence of a growing association between scientific research and technological innovation.[16] But the facts about mega-science suggest that a purely unitarian origin is unlikely. Science can be good and useful, also problematic, and mostly entirely irrelevant to practical affairs. Of more concern theoretically, though, is the surely faulty but popular logic that it is only and fully a societal *requirement*. Like an exuberantly welcomed new technological app, all science is widely assumed to progress towards useful ideas and things. Yet this is not so. The vast majority of science is for, and of, science alone, serving the elaboration and activities of future scientists and their theory-driven inquiry in a path-dependent process of feedback loops and increasing knowledge returns, until major breakthroughs restart the process. Contrary to persistent myths about science and society from both negative and positive viewpoints, a new product or societal solution or even a new crisis does not quasi-automatically result from each new discovery. And how could it be so, especially with the nearly incomprehensible volume of papers published annually?

Science generates new knowledge regardless of anticipated applications, funding, and science policy.

Articulated in a broadening range of disciplines and specialty fields, science moves over time to its own rhythms and by its own rules, and less so to some notion of the "needs" of society, whatever those may be at any given time. The natural sciences especially do not have one collective end in mind, neither the good society nor the evil society. The humanities and social sciences focus on such questions, as do distinctly nonscience activities—political, moral-religious, and so forth—that question the supposed implications of scientific discovery—and even guard the boundaries of scientific impact. Science, of course, is supported by a range of public and private interests and therefore often interacts with political, moral, and philosophical issues and forces within society, but it is not itself chiefly driven by these. For example, the world's science capacity was intrinsically motivated and financially supported with extraordinary interventions by governments and policymakers to respond specifically to the COVID-19 crisis. But scientists' successful efforts in vaccine development and therapies stood on a vast foundation of prior discovery whose initiation and elaboration was not necessarily connected to any specific societywide crisis, but rather grew from incremental improvement of medical knowledge and treatments. And last, science can be misinterpreted by nonscientific institutions, as evidenced by contemporary conspiracy theories and vaccine opponents in the public sphere up to the highest levels of government in various countries.

Parsons's 1970s ideas about the university and knowledge production are often criticized as being too optimistic about science and rationality. In contrast, the university-science argument here is not based on a notion of an objective, all-encompassing utility of science, although, as will be shown, a widening societal belief in exactly that, along with the education revolution, has certainly helped to amass the resources that made mega-science possible. Sociologists, including Parsons, have long recognized that a kind of "secular religion" of beliefs and assumptions about the rationalization of society is at the core of complex liberal society. As part of that, science has become a sweeping ideology that increasingly holds sway, regardless of whether it is objectively better or worse for society.[17]

The strength of the ideology, for example, is mirrored in the vitriolic opposition to scientization by illiberal positions in society.[18]

Depending on the context, although science can be beneficial or damaging, and certainly transforming, and there are at times specific applications, society does not necessarily "need" it, or at least not the overwhelming majority of it. Despite utilitarian calls for more scientific investment to serve societal goals, there is far less directed practicality behind science than is widely imagined. Sometimes, the most important discoveries solve a different problem than that currently in focus. Other times, the impact of an incremental improvement requires a window of opportunity provided by a different scientific subfield altogether. Even the largest, long-term collaborations in science, such as the CERN laboratory, astronomical observatories, and various supercomputers, may well require decades of investment without guaranteed results, and only much later might they confirm (or refute) a theory.[19] Popular versions of the same argument can also be put aside, such as science is "needed or required" by the military-industrial complex, growing social complexity, medical establishment, technological advance, political states, and so on. This is not to say that these and other parts of society do not welcome and employ scientific information—they all clearly do and will readily provide resources in exchange for it, all the while basking in the prestige of legitimate knowledge. But even though information derived from scientific inquiry is extensively employed, science itself does not function very well as a societal requisite since so much of its huge volume is functionally disconnected from anything other than science itself. In short, science has its own distinct culture.

Even the notion that science is self-perpetuating is a less powerful force than once assumed. Philosophy of science, itself well developed, continues to debate the logical structure of scientific theory and methods and how they might combine to propel discovery forward.[20] Earlier thought suggested a perpetuating linear progress from one finding and theory to the logical next. Instead, later convincing arguments found science to be a mixture of linear and nonlinear (less self-perpetuating) processes behind inquiry, including even a significant degree of randomness, or in the words of Robert K. Merton, a founder of the sociology of science, "serendipity."[21] Either way, philosophical thinking does not seriously take up the origins of the century of science, other than to suggest that whatever is behind science is obviously intensifying.

Other than the social utility, political, and economic lines of thought, a fair amount of historical description about science does add insights about what chiefly sustains it and might assist in the current explosion we are witnessing. First is that science as a method of inquiry evolved over a long time—many centuries—out of magic, religion, and myriad ways of knowing developed by earlier societies. In other words, it is not a new human activity, and recent intensifications build upon ancient roots. Second, science received a boost in status and volume during the sixteenth to eighteenth centuries in Western Europe,[22] a fact that created a near obsession with this period among past scholars of science, leaving later developments understudied. Clearly, though, the rapid progress of mega-science is not merely a progression from this earlier period—it is quantitatively and qualitatively distinct. Third, real people do science, so all sorts of human machinations, including social affinities, power politics, status competitions, and everyday pettiness and pride can sway or undermine it. Thus, of course, any explanation must include social interactions, but unless we think that the last dozen decades have witnessed a unique intensification in the root humanness of science, more is required. Last, as noted, there are self-sustaining qualities to science and to the values and ways of thinking engendered in each scientist, implying that causes of the century of science are likely those that enhance this enduring process. Yet that is about it. Other than to observe that it has become a major part of postindustrial society, the main drivers of the century of science were unclear. Essential inputs are evident: scientists; their socialization and *esprit de corps*; resources provided by governments, firms, and philanthropists; and reasonable working conditions, including the freedom to devote themselves to their own curiosity and to discovery without interference. Yet why these essential resources dramatically increased over the past century is not answered in the usual ideas about what grows science.

Mega-science is not only an abstract, technical, somewhat self-perpetuating, pursuit that is influenced by human interactions—it could not have occurred without a supporting culture and the eventual amassing of unprecedented resources from diverse sources committed to its advance.

As Parsons predicted, we will see evidence of a thoroughgoing confluence between the education revolution and science as an ideology and activity behind the fast-rising vast ocean of scientific papers. This is the jumping-off point to track how and why the university-science model became the main platform for global mega-science, a journey that starts with early clubs of discovery and moves on to an idyllic nineteenth-century German town that hosts what was then among the world's most celebrated universities.

CHAPTER 3

Göttingen and Beyond

The Ascendant German Research University

What Talcott Parsons so prophetically envisioned about universities and their major knowledge-science-producing role in world society began mostly in the German-speaking regions of Europe, well before the century of science. As already shown, the westward-moving center of gravity of scientific publications had reached approximately the middle of the northern Atlantic Ocean by 1900, reflecting a significant shift from European to U.S. universities examined in subsequent chapters. Its path also indicates that the trends of global mega-science originated farther east. In many respects, beginning as far back as at least the sixteenth century, seeds were already planted in Britain, Italy, France, smaller nations around them, and, of course, Germany. This seeding, in the form of institutionalizing new ways of doing science, increasingly was taken up by Germany's universities. And by the mid-nineteenth century, these were driven by an often-overlooked, emerging but modest growth in the centrality of the university in society that enabled growth of organized discovery and innovation.

Frequently, accounts of the earlier scientific revolution in parts of Europe focus on scientific breakthroughs of pioneering notables, such as Leibniz and Newton, and their discoveries that became key foundations of various disciplines: the popular image of a lone genius appearing out of nowhere to contribute a major discovery. A more accurate description, though, focuses on a rising culture that would come to scaffold scientific inquiry through common understandings among scientists. Accruing from about the end of the sixteenth century, this cultural scaffolding came to underpin and define the steadily maturing collectives of intellectuals attempting scientific inquiry. Once established, this culture would later lead to the core of the future university-science model, providing a platform for the development of new teaching methods, faculties, and, perhaps most transformative, a tightening interdependence across the often-divided missions of research and teaching. But it was not a simple transformation. The coming of a culture of science within the fledgling German research university was a gradual confluence of ideas about nothing less than epistemology, pedagogy, scientific inquiry, and the Realpolitik of empire building, the latter of which would ironically, over time, become a formable hindrance of greater articulation of the university-science model, leaving its further development to occur elsewhere.

THE DRAWING ROOMS OF SCIENCE

University science did not just naturally organize itself. Two historical innovations led to this path, one mostly extramural to the university, the other intramural.[1] The former was the rise of academies of science. By the early seventeenth century, external to universities, numerous local learned societies of gentlemen, often related to a particular city, were engaged in some semblance of scientific pursuit. Frequently, these were organized around private collections of specimens of anatomical, botanical, geological, and other such materials for study. Sometimes referring to themselves literally as "the cabinet" where the collection was kept, by the end of the seventeenth century these once salons of amateurs meeting in private drawing rooms were an emerging form of scientific collective. Initially eclectic groups of medical doctors, wealthy dabblers, government officials, and literary figures over the centuries gave way to a more formal membership of mostly professors across the then few established branches of the sciences. With subsequent royal chartering and funding, some became larger formal academies of science, such as the *Deutsche Akademie der Naturforscher Leopoldina*, founded in Schweinfurt, Germany, in 1652; the *Accademia del Cimento*, Florence, Italy, 1657; *The Royal Society*, London, England, 1660; the *Académie des sciences*, Paris, France 1666; and *The Imperial and Royal Academy of Brussels*, Belgium, 1772. From there, ties between these visible scholarly networks of academies and universities grew.

The academies were, crucially, the hothouses of a new culture nurturing debate over an emerging consensus around scientific methods, professional norms and rules, and, most important for our argument here, the public sharing of results in formal papers.[2] Written by the discovering scientists themselves, usually alone, these papers were published in forerunners to the current scientific journal, such as *Philosophical Transactions of the Royal Society*, appearing in London from 1665. Some standardization of methods and sharing of materials and results within a virtual community of scientists was a revolutionizing practice, referred to as an "invisible college."[3] But the academies enabled an even more salient change. While the historical record is complex across these centuries and countries, out of the countless conversations and formal meetings of academies emerged a newly accepted logic behind empirical research that was later assumed and further developed by universities. It was a gradual, broad epistemological shift away from Aristotelian, pre-Enlightenment logic of "received knowledge" (from God) to a new logic of "produced knowledge" (by contemporary humans). And with this, over a long course of time, methods of inquiry evolved from medieval disputation, whose ultimate goal was illustration of preconceived truths through choreographed argumentation, to the recognizable essence of contemporary science as open discovery based on systematic, empirical methods and written synthesis of findings.

This shift had far-reaching implications, changing everything about how science is produced, verified, and accrued. First is the deeper acceptance of establishing preference among empirical methods exclusively based on scientific logic. This was an important step on the long path to the widely accepted epistemological superiority of results from the controlled experiment. This change, likely begun by tenth-century mathematician Ibn al-Haytham during the middle of the Islamic Golden Age, culminated in the mid-twentieth century with advancing mathematics, which included probability theory and statistical estimation discoveries incorporated by developers of experimental designs.[4] Early attempts at methodological logic, no doubt facilitated by the self-evident validity of results themselves, grew in technical sophistication. Theory, systematic empirical observation, hypothesis testing, and eventually the value of replication of shared technical methods and results had roots within the early European culture of science. This collective shift in thought also yielded the pivotal idea that scientific discovery is not legitimately bound or limited by religious ideology nor, later, by the state or society. In other words, scientific inquiry was increasingly defined as epistemologically separate unto itself. Hence, any and every subject becomes legitimate for investigation. This autonomy and universality are now taken for granted. Yet we need only consider negative reactions to culturally controversial new scientific findings in an already highly scientized contemporary society—or indeed current political pressures relating to scientific advice and advance to cure pandemics and climate disasters—to appreciate how revolutionary this idea once was.

Gradually, metropolitan academies transformed into fully articulated national academies of science. Now a standard part of the state-craft package of liberal societies, these academies became more honorific, with their activities mostly limited to lending prestigious advocacy for a process of science that their forerunners had in part generated over the past centuries. The doing of science shifted away from academies, and by the turn of the nineteenth century, in a handful of countries, universities had become the primary site of research, following the rules and practice of an increasingly organized culture of inquiry. The stage for global mega-science was now set and, over the next 120 years, the plot would largely unfold within universities themselves.

We begin in Germany. The fact that the earliest research universities originated in German-speaking regions of Europe is well known.[5] And while the many details over a three-centuries path are documented in various German and English sources,[6] the full picture is not well understood, nor are the links to our contemporary global mega-science era sufficiently appreciated. Also, much is made about the research university model being subsequently copied in other nations, but the real story is at once subtler and more complex. It is not a nice, neat history with obvious causes, clear watershed events, or precise dates of initial trends. There was much resistance, inertia, halting change, even retrenchment; in short, anything but a consensual, unidirectional linear process. Also, interspersed among ideological shifts were changes in how universities actually operated and how specific ones faired long term.[7] And change in accompanying organizational behavior by universities was not necessarily seen as a sweeping change in the minds of the reformers themselves and could be misperceived in isolation as trivial. Indeed, the mid-nineteenth-century movement towards the research university was more of an incoming tide than a series of strikingly obvious events, so much so that the professors and occasional government officials facilitating them could hardly have predicted where they would lead in the near future. The tide, though, was a long time coming.

The evolution of the university in German regions had many twists and turns, and we cannot and do not need to cover all of them here, as we launch our journey late in the nineteenth century. As a brief background, at the beginning of the eighteenth century Germany's medieval universities were moribund, with hopelessly small enrollments taught by faculties offering an inflexible ancient scholasticism, and with fraying integration within society. It was not uncommon for leading intellectuals to proclaim them as irrelevant and in some cases advocate for their total disbandment, suggesting that scholarship and science could best be done in the academies. In many ways the first part of the eighteenth century was witnessing the slow death of the medieval version of one of the longest-standing institutions in Europe. And if it were not for their eventual revival, the capacity to undertake university-based science might never have occurred. Through a process influenced by the long geopolitical course of the Holy Roman Empire, the rise of sophisticated, educated state bureaucracies, and even some resistance to European democratic movements, over the next two hundred years through many travails and reforms universities came to a unique position in Germany.[8] Along with scientific research, secondary education and universities were by the middle of the nineteenth century a focus of state policy and resourced as nowhere else in the world. They also were increasingly connected to other institutions in German society, such as the Lutheran church, a widening elite, a new range of occupations, and, of course, the famed Prussian civil service.[9] On the eve of the twentieth century a distinct group of reformed universities were the forerunner of what would emerge worldwide over the new century. Their enactment of an ideology behind what we call the university-science model was widely considered as the pinnacle of a new melding together of educational and research missions within one organizational form. And, in part because of this, the late-nineteenth-century German university also demonstrated a coming greater integration within society that would bring both resources and a renewed cultural legitimation to higher education and scientific research.

"PACK YOUR SUITCASE AND TAKE
YOURSELF TO GÖTTINGEN!"

Instead of any one notable scientific or educational innovator involved in the rise of the nineteenth-century research university in Germany, perhaps the trials and triumphs of the Georgia Augusta University of Göttingen (founded 1737) best exemplify the rising tide of this rich historical story—as well as its contemporary challenges.[10] Certainly other newly established universities in the eighteenth century, such as at Halle (founded 1694) and Erlangen (1734), where the first reforms began, would undertake similar paths towards the modern university. After some resistance these would be joined by older ones at Heidelberg (1386) and Leipzig (1409), and, of course, the later established Friedrich Wilhelm University of Berlin (1809) at the heart of the Prussian Empire—today's Humboldt University of Berlin—and the newer University at Hanover (1831), dedicated to science and technology. Yet more than most, Göttingen, in a small medieval town at the geographical center of present-day Germany, symbolizes the successes and struggles of the European culture of science from the eighteenth century and its influence on the university-science model at the beginning of the century of science.

Initially a state library (founded 1734) and with an academy dedicated to inquiry in the spirit of the Enlightenment, the Georg-August-Universität Göttingen was founded, intellectually and strategically, on a devotion to the free exchange of ideas without censorship. Unlike in typical universities of the day, the Faculty of Theology was not given a dominant position over the other Faculties to avoid theological censoring among the university's faculty, and to gain attraction among the high-tuition-paying young nobles as well as recruitment of faculty with Protestant and Catholic theologies. Alongside the university, the Göttingen Academy of Sciences and Humanities (1751) provided resources and platforms for interdisciplinary dialogue and a strengthening of the faculty of philosophy, the forerunner to the eventual inclusion of scholarship on all other topics besides the classical curriculum of just theology, law, and medicine. Scholars of worldwide repute were honored (and handsomely compensated) to receive a call to become professors in Göttingen. Among them were Albrecht von Haller, a physician, naturalist, and poet; Georg Christoph Lichtenberg, a physicist, philosopher, and author; and Carl Friedrich Gauß, a mathematician and astronomer. An important chapter in university history occurred a hundred years after its founding when King Ernst August of Hanover decided to rescind the constitution. Seven professors of the Georgia Augusta, notably the brothers Jacob and Wilhelm Grimm, philologists, lexicographers, and authors, submitted a written protest. The "Göttingen Seven," in standing up for the constitution and liberty, lost their positions and some were banned from the kingdom, yet Göttingen survived and eventually thrived.

Göttingen enjoyed a growing internationally celebrated academic reputation over the course of the nineteenth century, by the public and many of the world's most accomplished scientists, later solidified by its Nobel Prize winners, especially in the early twentieth century.[11] When a mid-century generation of aspiring American scientists and other intellectuals sought out German universities for training, the University of Göttingen was often first on their list, akin to being atop the international rankings of the world's universities today. Initially a saying among mathematicians, the exhortation to immediately get oneself to Göttingen came to reflect a general belief that something new and special was happening there.[12] As a hub of cutting-edge science, the town named its streets after famous professors even before their retirement, placed their headshots on postcards for tourists, and publicly traded quaint stories about eccentric intellectual habits and accomplishments, reflecting a kind of academic idyllic setting within a picturesque town in Lower Saxony. Although this was no doubt partly the kind of hype bestowed upon any university that succeeds in being internationally recognized as among the world's best of an era, along with others such as the universities in Berlin and Hanover, the University of Göttingen in what has been called its "golden age" had become the prototype of a reformed model for a university.

Of course, the long history of institutionalizing science within universities belies the notion that the late-nineteenth-century research university scene sprung wholly anew. And seemingly like every major cultural change in Germany, Göttingen and other emerging research universities were an articulation between philosophical change and Realpolitik—idealism fused with the pursuit of political interests—usually taken to an extreme as was customary in German culture.[13] In this case it was Bismarck's type of Realpolitik, emerging from the relatively late development of the German Empire led by the Prussian state. On the cusp of surpassing Great Britain's world-leading steel production, and with its 1870 victory in the Franco-Prussian War, the empire was defining itself as a new world leader. An exuberance of triumph fueled further advanced educational development on top of Prussia's historically early provision of state-supported, compulsory mass primary and secondary schooling for a smaller, elite portion of youth. The newly unified German state significantly increased resources provided to administer universities along with individual provincial administrations. This arrangement, under which the newer University of Berlin—in effect the Prussian Empire's attempt at a pinnacle university in its capital—did very well, as did many other research universities. And so eventually would Göttingen after initially suffering some under this new political order because of the long history of political competition between Prussia and Hanover, including the latter's long-standing connection to Britain's Royal Society and earlier alliance with the British Crown, considered one of ascending Prussia's main competitors. But despite lower allocations of resources and intermittent state bureaucratic interference, Göttingen's innovations, like those at other top universities, were robust enough to steer it into a golden era.

Underlying these innovative practices were new ideas, ideals really, that were brewing in a number of universities in Germany and became a set of values facilitating the university-science model. What are widely acknowledged as the nineteenth-century "Humboldtian" ideals behind the research university stem from four principles: academic freedom to teach, learn, and research; unity of teaching and research for deeper inquiry; broadest possible inquiry including science and humanities; and the primacy of "pure" or basic science in striving for universal knowledge. All are underpinned by the elusive, yet frequently extolled, notion that participation in advanced education and its scholarship, and by extension scientific discovery, is essential for nothing short of full human cultivation (*Bildung*).[14] These clearly fit with, if not also partially originated from, the spirit of the initial institutionalizing of the culture of science within the earlier academies. Despite their centrality as guiding principles, contrary to popular belief these never represented a formally articulated organizational blueprint for universities in Germany or elsewhere, nor did they mostly spring from their widely assumed originator and eponym, the humanist Wilhelm von Humboldt (who as a young scholar studied at Göttingen late in the eighteenth century), nor were they part of one concisely derived model arising from a single university at a single point, such as is often assumed about the University of Berlin (renamed from 1949 after the Humboldt brothers Wilhelm and Alexander, the explorer). Of course, the aforementioned provincial and central ministerial administrations that partially governed universities did, more or less, support these ideals and did operationalize them within the administration of higher education in various ways. And by the middle of the nineteenth century these ideals were spreading out from Germany to the U.S., with new modifications on them flowing back to German universities.[15] Nevertheless, over a long developmental period, influenced by many thinkers and educationalists at a number of universities primarily in German-speaking regions, there was enough appeal of these principles to guide more classical universities into becoming invested in research.

The now taken-for-granted principle of academic freedom to inquire about any subject had a long and complicated development, starting in the earliest medieval universities. For those most accustomed to the American university, winning freedom to expand or create new disciplines, curricula, degrees, and research areas may sound strange. But initially through tradition, later reinforced by various forms of higher education governance, European universities did not have the unquestioned charter to organize the study of novel topics—natural sciences, social sciences, and business being examples of latecomer disciplines. Later in the eighteenth century the earliest reforms towards greater intellectual freedom saw Göttingen and other universities including courses and scholarship on what were then newcomer academic topics of history, philosophy, and new legal training, all of which became valued by securing additional sources of students. And German universities would continue to take full advantage of academic freedom. Hence, by the 1890s, Göttingen had established a series of independent departments of mathematics, astronomy, physics, chemistry, technology, and mechanics, in effect emphasizing and resourcing natural sciences as never before.

Broadest possible inquiry and primacy of "pure" or basic science for universal knowledge is an ideal that also did not occur rapidly. Nor did it have a simple linear path of development. In fact, there was rather complex cross-pollination between Göttingen and some American universities over the incorporation of theoretical science and engineering in the first decade of the twentieth century. But the central issue here is that as a budding research university, Göttingen was propelled by, and in turn accelerated, broad scientization. Thus it was at the heart of the embodiment of the sweeping trend of rationalization in modern society, of which science is a central ingredient, observed and written about so keenly by the German sociologist Max Weber while himself a professor at Heidelberg during this exciting period of first modern research universities.[16] For example, being at the world's center of mathematical research at the time, Göttingen facilitated the development of mathematics as the universalizing language of science that would have such an expansive influence on the scope of science.

So too, the oft-celebrated ideal of unity of teaching and research was embodied in new practices of interaction between faculty and students. At least by 1800, professors at Göttingen and Halle, and then soon after extensively at Berlin too, were using a new form of instruction that combined teaching and research in the now familiar "academic seminar."[17] An innovation that accompanied greater institutionalization of empirical research, the seminar combined experienced faculty-scientists with students to read, discuss, and critique ongoing lines of scientific inquiry and theory. Now ubiquitous in universities worldwide, the seminar was a major improvement in training and collectivizing research. In contrast to the traditional one-way lecture or one-on-one tutoring famously used at Oxford and Cambridge, the seminar not only integrates more students into research but also influences faculties' theoretical development and motivates future research. In addition to the benefits from the seminar, an acknowledgement of the reinforcing process between any form of teaching and an openness to ideas for new scholarship and research steadily developed. So much so, for example, that a contemporary, prominent, university-based philosopher of physics can attribute teaching as an inspiration for new scientific theory: "When you teach, you're forced to confront your own ignorance. . . . [W]hen I was teaching a class on space and time . . . I realized that I didn't understand it. I couldn't see the connection between the technical machinery and the concepts that I was using."[18] This reflection would have seemed strange for an eighteenth-century professor. Later at Göttingen, as elsewhere, would come greater organizational integration of training and research evidenced by university-based scientific and medical institutes with their own lecture halls, teaching laboratories, and an organization of scientific labor to accomplish research efficiently. Indeed, a vice president for research from any contemporary research university in the world would fully recognize and endorse the logic behind the instructional and scientific research practices already in place at Göttingen and other similarly reformed universities by the late nineteenth century, while the reigning form of the more classical university just a century before would strike the V.P. as an alien place for scientific research.

The emerging university-science model, essentially Humboldtian ideals enacted organizationally, was not lost on Göttingen's professors and their sense of important professional activities. Observers of the university's idyllic academic setting often commented on a distinct undercurrent of competition and pressure among faculty in the sciences and emerging social sciences to produce scientific research and publish papers, so that one new faculty member likened the competition to a near-suicidal work atmosphere. This was the beginning of a "publish or perish" mentality at universities that has pervaded the mega-science model from its earliest stirrings. Prior to the rise of the research university, although scholarship was an essential activity, many faculty did not regularly produce scholarly products, at least not as an ongoing salient universal requirement of their positions, and there were also nonscholars with university posts for political and elite cultural purposes. Relatedly, while Göttingen resisted the Prussian state's attempt to shift the most talented young academics to the University of Berlin in what was referred to as the "Berlin system" of faculty development, it nevertheless was an early participant in the freedom of movement of faculty and hiring of talent by universities, a forerunner to the now global "star system" of faculty recruitment that runs on evidence of quantity and quality of scientific output. The same new professor who noted the extremely competitive environment also acknowledged that there was no better place for a scientific career than at Göttingen. While not explicitly part of the Humboldtian model, publish-or-perish and a star logic to the faculty recruitment are in fact clear manifestations of the ideals' overarching focus on incorporation of scientific inquiry within universities on the eve of the twentieth century. In dreamy Göttingen, it was the daily enactment of these behind-the-scenes ideas about combining scientific research and the university that led to celebratory academic street names, legends of high-pressure scholarly achievement, and exhortations to attend in person if one had the means for such mobility. In contemporary times, Göttingen has struggled to maintain that same global preeminence it once enjoyed, yet despite recent setbacks associated with the Excellence Initiative, the university remains among the best in Germany.

EARLY IMPACT OF THE EDUCATION REVOLUTION

Enactment of Humboldtian ideals would have meant far less to the growth of science had they not been accompanied by a shift in the relationship between education and society. Yet with all the scholarly focus on the Humboldtian tradition, the dynamics of Germany's early maturing education revolution, slowed as it eventually was, have rarely been acknowledged. It is both the articulation of these ideals *and* a widening cultural centrality of universities that provided fertile ground for the development and diffusion of the university-science model so evident in Göttingen.

Even though German-speaking regions of Europe were early to make strides towards mass primary schooling and incorporation into secondary education, for most of the late eighteenth and early nineteenth centuries only a minuscule proportion of young men, about 2 to 5 percent, had attended a university, and then primarily for preparation to become ministers, physicians, attorneys, high government officials, and, of course, new professors.[19] After the 1871 unification of the German Empire with its elaborate Prussian administration and even within a tightly stratified education system, there was an increasing inclusion from the middle ranks of German society into the university. Now, the sons of not only the feudal elites, but also of the bourgeoisie, large farm owners, technicians, artisans, and merchants attended, often being the first among their extended families to do so. From 1870 to 1930, university enrollment for males grew at a rate faster than in France and the U.S., particularly if only American universities, not undergraduate colleges, are included. Early in the nineteenth century, approximately one-half of German university students had a father with a university degree. With rapid subsequent growth in enrollment this group proportionally dropped by one-half, as new portions of the population were undertaking advanced education for the first time. This was accompanied by a broadening of the cultural meaning of the university's authoritative training and degree-granting charter in German society, including in its labor market for positions in large-scale firms and their technical production, so that advanced education became a significant part of social stratification. This process was subsequently reinforced by the growing knowledge production in universities.

These cultural currents were reflected in what these new types of students studied once enrolled and what their degree meant once completed. As would happen repeatedly worldwide over the coming twentieth century, the classical orientation of the university would recede as natural sciences and social sciences grew. Across Germany in 1887, for example, over 25 percent of university students were studying theology, 19 percent law, 33 percent medicine. Just twenty-five years later, theology had dropped to 9 percent and medicine to 19 percent, while law remained about the same. Over the same period and against a backdrop of overall growing enrollments, students majoring in the humanities, natural sciences, and social sciences doubled from less than a quarter to almost half of all students. And among these, about 40 percent enrolled in natural sciences and social sciences. University degrees progressively became requirements (*Berechtigungen*) for positions in the expanding upper civil service of experts. The traditional elite also steadily added a degree from a university to its symbols of power and privilege. So too, a growing industrial economy requiring higher-skilled R&D and technical positions hired university graduates as they never had before. By the fin-de-siècle, an estimated one-fifth of young men had completed the upper secondary Gymnasium and attained the *Abitur* (the examination required for university admission) and were attending university, including at the more technical colleges, *Fachhochschulen*, along with a smaller, but steadily increasing, number of women.

Back in Göttingen, the university was already enjoying the rise in university training. By 1866 it had thirty-two professors (akin to department heads and laboratory directors supervising many graduate students and young PhDs in research) for mathematics, astronomy, physics, chemistry, botany, zoology, mineralogy, geography, and medicine. It was the largest such natural sciences faculty among all German universities and over twice as large as its future rival in Berlin. Even though the University of Berlin would receive the lion's share of resources over the next few decades, Göttingen was enabled by growing education budgets, enrollments, and cultural legitimation to expand its already substantial natural sciences training and research into more specialized, cross-disciplinary areas, a hallmark of megascience. In 1896 it established a new professorship in physical chemistry and electrochemistry, another in geophysics and applied mathematics in 1904, and one each in applied mechanics and applied electronics in 1907, the latter of which would head the first university research group on this topic in the world. A sense of the new empire, a rising tide of Humboldtian ideas and an aspiring university culture of knowledge production, and an advancing education revolution would spread the university-science model across more universities than in any other nation at the time.

On the eve of the twentieth century and well into Göttingen's golden era, universities across Germany had undergone a long, slow sea change of mission and form in who attended and how central the institution had become within the society. These changes were the forerunners to the very trends that stimulated Parsons's observations seventy years later. It is not surprising then that by 1900 Germany already had approximately thirty research universities whose faculties were publishing STEM+ papers in a growing set of journals that, although at first sponsored by university departments themselves, would come to operate much like the contemporary scientific journal, often written in German, the scientific lingua franca at the time. This was the largest set of research universities in a single country. The United States, for example, with 25 percent more population, would not supersede this many research universities until the 1920s. Therefore, Germany entered the century of science with the greatest number of functioning research universities, which enjoyed enough academic freedom and integration of education and research to fashion significant capacity to generate new science.

Importantly, Germany was not alone in this early process. Over the first three decades of the twentieth century, the most active scientists were in Germany, the U.K., France, and the U.S.[20] These four countries had begun to increase educational opportunities for successive generations of children and youth, particularly from the late nineteenth century on. The education revolution gradually but steadily flowed more students into already established universities, supporting and enlarging a faculty that progressively undertook research as a central part of their position and mission.

"KNOWLEDGE FACTORIES!"

Their innovations were eye-catching enough that Göttingen and similar universities of the period were pejoratively likened to "knowledge factories," a much-repeated criticism then and still now by proponents of classical curriculum in response to declining legitimacy of older models for the university over the ensuing century. And although in form, factories they were not, producers of unprecedented levels of new knowledge they most decidedly were.

Germany and its universities were at the center of the nascent pattern of global mega-science: out of an estimated sixty-three hundred STEM+ papers from all European countries published in 1900, 70 percent included at least one Germany-based scientist, making it the largest STEM+ article producer in the world at the time. And as shown in Figure 3.1, at the start of the century of science, alone smaller Germany even outpaced American scientists, the future world leaders. While papers from the latter began to eclipse German production a decade later, German universities continued to turn out a significant number and, like the U.S., experienced an increasing growth rate in the 1920s. And as shown before, along with Germany, university-based scientists in other European countries made the whole region responsible for the largest volume of STEM+ papers in the world up until about mid-century.

As Figure 3.2 further demonstrates, the top five most productive German universities published more STEM+ articles in 1900 than their counterparts in the U.S.; from 1910 through 1920 their output was approximately similar. By 1930 though, the top American universities had begun to pull away, likely, as examined in the next chapter, because of their unplanned

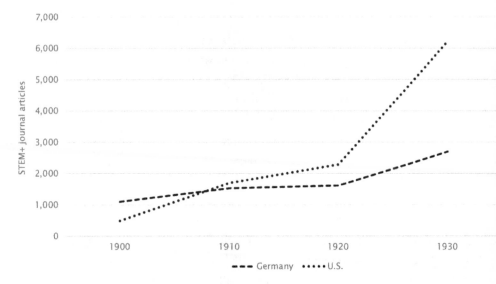

Figure 3.1. STEM+ Journal Articles by German and U.S. Scientists, 1900–1930. Source: SPHERE database. Note: Research articles with at least one author based in each country. In these decades, very few papers were coauthored.

innovation on the university-science model and a deepening centrality of the university in society. What is important about this shift is not a race between German and American universities; rather it is that by 1900 a large share of the world's scientific publications was being produced by a few dozen universities whose research mission and general organization would eventually be copied, transcending national borders. While, of course, it would also take parallel economic and political developments to spread, and at times some national traditions in higher education worked against the model's adaptation in countries, the well-known transcending influence of the education revolution would carry along the university-science model. Also, well before American scientists, faculty members at German universities were starting to collaborate and publish with scientists at universities in other countries, a pattern that would become more pronounced over the century. While things occur at places and times for a range of specific reasons, they are often tied to less obvious, larger changes that are taken for granted. From its beginning global mega-science was more about collectivizing science at the university and its centralizing relationship to society than about any one nation's efforts, culture, or geopolitics.

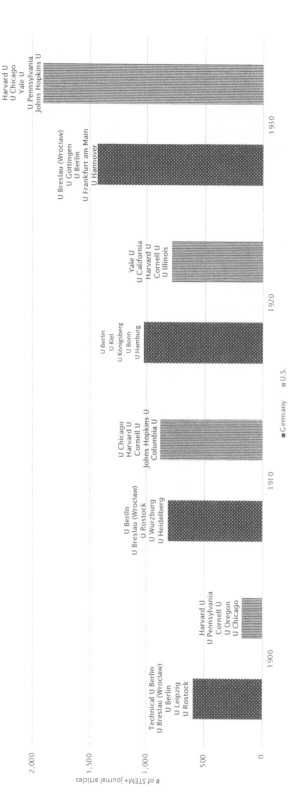

Figure 3.2. Most Scientifically Productive Universities in Germany and U.S., 1900–1930. Source: SPHERE database. Notes: Estimates of approximately the amount of STEM+ publishing by each university derived from a sample of journals; this does not represent an actual census of all publications. Not included in Germany's lists are several highly productive extra-university research institutes (see Chapter 6 for further details on the two pillars of Germany's science production).

On the eve of the First World War, the University of Göttingen had prevailed through the Prussian political favoring of the University of Berlin, with many faculty seeking to be there. Göttingen had the largest, and putatively the best, natural sciences faculty with forty-eight professors, each directing laboratories staffed by younger PhD scientists and graduate students. Crucially, the university was also enrolling many undergraduates in the natural sciences, ranking second in the world only after the University of Berlin. Consequently, its estimated STEM+ paper production consistently put the university among the top ten most productive in the country, and by 1930, during a significant boost in the overall publishing rate, it was among the top ten of the most productive universities in the world. Göttingen, like other universities in Germany, expanded new scientific discovery through a successfully articulated version of the Humboldtian ideas while riding the slow rising tide of the basic ingredients of the education revolution. These first stirrings of modern scientific research in the university would become the prototype for the future. The future, though, would arrive rapidly, for mega-science was in transnational motion exactly at a time when Germany began to turn against its own highly productive universities for ideological reasons. Not surprisingly, the Nazi regime and World War II along with world economic depression in the 1930s would shatter Germany's preeminence in science. From 1931 through to 1939, more than three thousand scientists—approximately two-fifths of the total in German universities—were forced to leave their positions because of their Jewish heritage or oppositional political affiliation, with a significant number emigrating to relatively familiar university settings in the U.S. Göttingen, for example, lost a fifth of its faculty, including some of its most distinguished scientists, while the jewel in the crown, the national university in Berlin, lost almost a third. Besides losing valuable scientists, those who remained lost collaborators worldwide, a further cause of reduced productivity.[21]

Understandably, much is made of the tragic purges of scientists and other intellectuals; from 1933 on the National Socialists also rejected key parts of the Humboldtian ideology and dismantled the university-science model with the support of a majority of remaining faculty and much of the student body under the grip of nationalist fever.[22] As sad as this latter period would be for the Germany university, ironically it further demonstrates exactly those salient features of the research university that would be repugnant to an illiberal, undemocratic, and, to a degree, anticapitalist regime, thereby reinforcing Parsons's insight about the compatibility among democracy, capitalism, and the education revolution. For example, the very essence of *Bildung*, the assumed moral ennoblement from scholarship, including scientific research, was rejected as too apolitical. The regime wanted an explicitly "political university" as the elite supreme school of, and for, the state, not one based on research and knowledge creation, administered and funded by the state but also self-governed to a degree by academics. There also were many attempts to reject scientific facts and theories and to install in universities a kind of pseudoscience and false intellectualism based on the party's core notions of radical racism, totalitarian nationalism, and a refutation of impartial experts, plus a downgrading—or worse—of former cultural elites. Unconditional scholarship and academic freedom in teaching at the heart of the Humboldtian ideal were deemed as obsolete, unscientific ideas. So too, the growing centrality of the university, in part reinforced by advanced capitalism's growing requirement for trained experts, was curtailed as enrollment was restricted and then denied altogether to women, Jews, those judged political and social undesirables, and an ending was made to the centuries-long tradition of training foreign scientists and scholars. And perhaps the National Socialists's most brutal indication of their opposition was, upon military occupation, the closing of all Polish and Czechoslovakian universities and interning of several hundred of their faculty in concentration camps.

By the 1940s, the devastation of human and capital resources plus ideological purges and the collusion to undo the Humboldtian ideology and deactivate the university-science model rendered only twenty-four German universities—with weakened capability—still publishing basic science in the world's leading journals. At their zenith in the 1920s universities in Germany were publishing 60 percent of all European STEM+ papers while drawing ever more youth to their doors. But by 1940 Germany's contribution to a continually expanding flow of European publications had dropped to just over 20 percent and would eventually fall to below 10 percent by 1970.

Even international acclaim could not save the University of Göttingen from the same fate. After the 1920s, when eight of Göttingen's professors in physics and chemistry became Nobel Laureates, its intellectual and research atmosphere steadily deteriorated over the late 1930s and into the 1940s. Particularly decimated was its core cluster of world-class mathematicians, many of whom were Jewish. It would not be until the late 1960s that Göttingen's faculty would be revived by a new wave of enrollment and along with other West German universities would resume their former levels of productivity, signaling the postwar rebounding of higher education and science systems across Europe. It is remarkable that despite the tragic European turmoil from the 1930s until the 1950s, the university-science model survived, as it could have easily vanished among the nationalistic chaos. One reason was its appeal in general and another was that Humboldtian ideals and initial implementations moved with the steadily shifting center of gravity of science towards North America. The U.S.'s unchecked public and private growth in universities took the lead in the world's unfolding education revolution—and with it a dominance in research that would last far into the twenty-first century. Next, we turn to how the U.S. achieved this feat from modest origins.

CHAPTER 4

Modest Origins

The Expansive American University-Science Model

Even before the decline and eventual devastation of German research-intensive universities, the world's center of scientific paper production had steadily traveled towards the American shore. Scientists in the U.S. were already authoring about a fifth of the world's STEM+ papers in 1900, a rate that would soon explode so that by 1930 they were responsible for almost half of global mega-science. At the rudder of this westerly shift was the university-science model that had been hatched in Europe and grown into the prototype of the research university at the heart of mega-science. The scientific excitement at Göttingen and other continental universities was not lost on a few Americans who hoped to develop similar universities. But like all sociological forms that guide complex organizations, what was traveling over the ocean was not a rigid blueprint of how to form a more research-intensive university. Rather, it was a loose combination of guiding ideas, now somewhat mislabeled as "Humboldtian" ideals, and an early record of successful research fueling an infectious aspiration to emulate this European organizational form in the U.S. Nor did influence only flow west, as early ideas about science and universities from the U.S. were incorporated into subsequent developments in Europe.[1] The general story of this trans-Atlantic borrowing is well known, even legendary, in historical accounts of prestigious and scientifically productive American universities such as Johns Hopkins University and the University of Chicago, Harvard University, and the University of Michigan. These universities' leaders were influenced by their experiences of and ideas about the German university and became key reformers. These and several others, such as Clark University, were intended to be copies of the European research-intensive university of the time. What is more telling than a few imitations, though, are the distinctive ways in which the university-science model matured and intensified in the U.S. across a wide range of universities,[2] and how much the country's accelerating education revolution played a central role in this process.

When experts rank the world's universities by overall quality, including science and other scholarship, consistently U.S. universities make up almost all of the world's top twenty, and over one-half of the world's top one hundred. This is about four times more than the ranked universities from second place United Kingdom and third place Germany. And as noted before, even as mega-science continues to expand, scientists researching in the U.S. and Canada are involved in approximately a quarter of all annual papers. Because of the U.S.'s outsized role in the development of mega-science, its highly prestigious, wealthy, and often private universities receive a lot of attention in attempts to explain the country's contributions to mega-science. Yet these are only one part of the story, and in some ways not the main part. Earlier considerations of the American research university place much emphasis on the best-known ones among them, focusing on their wealth in multibillion-dollar endowments and expensive, star-studded faculty. And understandably, their unique resources and world-class reputation deserve attention in explaining the U.S.'s science production; yet it took far more than a handful of universities and a process more complicated than raw economic advantage.

As already evident in the German case, the early education revolution's cultural support of a spreading university-science model became a foundational, proximal, and enduring factor in the undertaking of large amounts of research. Nevertheless, in Germany and other parts of Northern Europe universities were still heavily wedded to their traditional roles of the preparation of societal leaders, elites, and professionals. If not for further development of the cultural and organizational manifestations of education growth in the U.S., the platform for global mega-science would have been greatly limited. In partially employing earlier innovations, partially misinterpreting them, and partially resisting a native nineteenth-century ambivalence towards advanced education, the Americans evolved their higher education system along a more radical path. With no concerted national plan, policy, or strategy, the American culture of higher education and its role in science intensified and consolidated innovations to the university-science model, lending enough support for scientific research with enough flexibility and crucial grassroots resources to lift the nation's capacity for science to a level unimaginable in the nineteenth century. And then these changes were turbo-charged in such a way as to make an expanded university-science model the main platform for the twentieth-century unfolding of global mega-science.[3]

Unlike European nations and elsewhere, the U.S. federal government does not directly control the supply of undergraduate higher education nor graduate training. After World War II, it did support and fund both in central respects, and there has always been interuniversity competition for national, state, and local resources, but the founding and chartering of universities with sprawling graduate programs were neither tightly nor centrally controlled. The crucial distinction between what started in Europe and what would occur in the U.S. is that the latter was the first country in which the ideas behind the education revolution fully materialized. In other words, the key factors behind the success of the American model are a certain degree of openness, accessibility, and democratization of advanced education blended into the production of science.

Without a controlling central federal administration, the country's nineteenth-century tradition of local schooling enabled the education revolution to blow through American culture like a wildfire, deepening the acceptance and the realization of mass education beyond primary schooling. The organization of education in the U.S. certainly has some widely noted problems at all levels, particularly social and geographical unevenness in resources, quality, and opportunities. Nevertheless, the U.S. has been at the leading edge of the education revolution in expanding education and weaving its values into everyday life. As previously noted, it was among the first countries to strive towards primary education for most children, achieving near universal, comprehensive secondary education by the 1960s. And, crucially, it never relied on streaming a high proportion of youth into vocational, nonuniversity training, unlike many other systems of education around the world, particularly in Europe and those colonized by it. As shown in Figure 2.2 in Chapter 2, the chiefly comprehensive system of secondary education expanded after World War II, with more Americans rapidly pursuing higher education at a variety of organizational forms, including community colleges, liberal arts colleges, public regional colleges, land-grant universities, and private universities. By the second decade of the twenty-first century, over one-third of all adults had earned at least a bachelor's degree, an increase of over twenty percentage points since 1970. In the span of a few decades, significant access to higher education occurred, so that at present, upwards of 70 percent of a national cohort of young adults are attending a wide variety of two-year and four-year colleges and universities. And as will be shown later, an even more extreme expansion of higher education provides the platform for major science production in Japan, South Korea, and Taiwan, and likely in China as well.

The supply of higher education to meet growing demand was, and still is, largely open and without restrictive government charters on the founding and development of universities, with enactment of their self-determined missions, including how research-intensive they could compete to be. This process can be summarized as a *symbiosis of access* to mass undergraduate education, decentralized founding of universities, and flexible mission charters for PhD training. Many youth had the access to participate in mass undergraduate education; decentralized actors had the access to establish universities; and universities had the access to compete in doctoral education and research through self-determined changeable mission charters. Further, these sources of participation influenced one another symbiotically by feeding mutually beneficial relationships. Along with other resources central to science, this access symbiosis yielded significant growth in new scientists researching at an increasing number of public and private research universities that continue to facilitate the nation's significant science production. Over the twentieth century, the American undergraduate enrollment rate and number of universities with STEM+ graduate doctoral programs each doubled three times and the annual volume of newly trained PhD scientists doubled six times. Generating unprecedented research capacity in this way enabled the U.S. to surpass early European science production, remain dominant for the rest of the century, develop a robust university-science model that would spread to other countries, and be at the heart of the steady globalization of science. With so much of the ideology of the education revolution intertwined into everyday life, including universities, American culture harnessed the motivations for advanced education to its development of universities and enactments of a robust university-science model to spread a highly effective platform for science: a large and highly differentiated set of postsecondary organizational forms, all increasingly participating in producing American research capacity.

Three stories, one here and two in the next chapter, illustrate how the interaction between access symbiosis and the university-science model had consequences for the growth of mega-science. First is a historical sketch of the hard-won cultural roots for the land-grant, public American university that, once applied and repeated through various local circumstances across the country, combined interests within the education revolution to open the way for greater numbers of research-active universities. In contrast to an older tradition behind a small set of elite, highly productive, mostly private universities, the newer cultural mix is a major engine behind the country's scientific successes. Subsequent analyses substantiate this process and show how the organizational implementation of the American university-science model yields exceptional capacity for research at a range of universities.

EVAN PUGH'S DREAM FOR A
RESEARCH UNIVERSITY

The oft-told late-nineteenth-century tale goes as follows. Small-time farmers relished pulling up their one-horse wagons in front of a fancy new agricultural experiment station at some fledgling local American land-grant university and asking one of the scientists how much profit their acclaimed "model farm" made last year. Upon hearing the anticipated reply that the station did not turn a profit and indeed required considerable additional funds to run it, the chortling farmers would smugly click their rigs down the road.

Likely a joke, this story nevertheless reflects the deep-rooted resistance across nineteenth-century America towards the university, science, and its application to everyday life, a culture that hardly appeared on the verge of world dominance in the creation of new science through an expansive version of the research-intensive university, although it was precisely that. Why the joke's sentiment was so popular; why so many of those universities were often in the middle of nowhere, publicly funded, and considered local and non-elite; plus why their agricultural experiment stations were forerunners to a new take on the university-science model have been well documented by historians of American higher education, so the full account need not be repeated.[4] Rather, what is important here is a sociological interpretation of how a growing number of public universities, including the land-grant universities and new private ones, joined into science production through the trends of the access symbiosis that would initiate and strengthen the cultural forces that Talcott Parsons so astutely observed by the 1970s.

Fittingly, the eventual domination of science by the American version of the twentieth-century university began with nothing less grandiose than manure, actually "artificial manure," as fertilizer was called at the time. By the middle of the nineteenth century, arable land east of the Mississippi River was discernibly diminishing, and the much-touted problem of overfarmed, exhausted soil loomed as a personal, communal, and even national economic threat, likened dramatically in newspapers to what had supposedly caused the fall of none other than the Roman Empire. Playing on this fear, unscrupulous companies offered farmers free, but bogus, chemical soil analysis that unsurprisingly always recommended the company's own brand of artificial manure—wildly overpriced and often ineffective. This otherwise pedestrian episode in American agricultural history galvanized an array of interests in bringing forth the public research university, forming a culture of educational access, diversification, and a democratizing of science.

The life and times of one Evan Pugh (pronounced "pew"), 1828–1864, epitomizes this process.[5] Who he was, the struggles he faced in establishing an early public, local college for future farmers, and how his vision of integrating research with undergraduate training went into the land-grant model upon which eventually highly scientifically productive universities spread across the country demonstrate the cultural roots of the American story. Pugh chiefly played a role at the local, not the national level, which emphasizes the larger point: the stage upon which the Americanized university-science model developed was not centralized. It was nationwide, but there was little in the way of a centralized administrative plan, or even strategy. Pugh was central in laying the foundation for just one public university. But this one case's origins, development, future training of new generations of U.S.-based scientists, and unprecedented production of new science epitomize how the Americanized model repeatedly unfolded in dizzyingly varied local ways across the country.

As postsecondary education grew in the second half of the nineteenth century, a new, if somewhat amorphous, educational culture was stirring nationwide. An unmistakable restlessness with the classical curriculum of European origins alongside a yearning for a new model of advanced education is reflected in one early reformer's strange battle cry—"the Differential and Integral Calculus will commingle with the ring of the anvil and the whir of the machine shop."[6] And Pugh personified this unique blend of cultural qualities driving these new ideas. He was a son of five generations of Welsh-Quaker Pennsylvanian blacksmiths and thus hardly part of the mid-nineteenth-century elite. And at over six feet, robustly athletic with piercing dark eyes set in a squared forthright face, he was hardly the picture of a scientific nerd. He knew farming and its challenges firsthand, including fraudulent fertilizer, and later in his life he could still vigorously out-plow younger male students in the research and training fields. Pugh was educated and had a talent for applying the emerging new science about chemical components of soil, authoritatively developed by Justus von Liebig at Gießen University in Germany (now named Justus-Liebig-Universität Gießen) in the 1840s. Like many of his American scientific peers, Pugh pursued the PhD at a German university. Aspiring American scientists were acutely aware that German universities were world-renowned and that something excitingly new in science was occurring at them. Pugh studied at the universities of Leipzig and then famed Göttingen, where he was awarded the PhD in chemistry in 1856. He then undertook a kind of postdoctoral training at Britain's Rothamsted agricultural experimental station, where Ronald Fisher would some seventy years later show how effectively to couple experimental design with statistical methods, thereby revolutionizing the ability to test scientific hypotheses and propel the computational basis of science.[7] Pugh also published an influential paper on nitrogen fixation in the storied journal *Phil. Transactions*, and presented it to the Royal Society with a large canvas diagram of the findings.

In short, from rather humble beginnings, Pugh took a path to a STEM+

research career fully recognizable a century and a half later: the PhD at a research university, followed by postdoctoral research, leading to journal paper publishing, right down to presentations at scientific conferences aided by the nineteenth-century version of the now ubiquitous Power-Point display. Notably, had Pugh been a European from his family's social position, it is doubtful that he could have taken a similar education and scientific career path, nor could he have participated in university building to establish a new version of the university-science model, as he did upon his return to the U.S.

Pugh had a big dream of creating the best agricultural college in the nation, then turning that into a system of colleges. This was a vision forged through his European experience, yet its audaciousness was pure American nineteenth-century aspiration. At that time the country was just open enough that such a dream was not out of the question, at least when approached one college-university at a time. Indeed, in what would now be an impossible career start for a newly credentialed PhD, Pugh sought and was appointed to the presidency of the modestly named Farmer's High School. Under his skillful leadership, it would soon be incorporated as the land-grant Pennsylvania State University (now known widely as Penn State). Penn State would eventually develop into a large public research university, and along with other early public institutions such as the universities of Wisconsin, Illinois, and Michigan, would be joined by many others over the ensuing decades. An untimely death at thirty-six cut short Pugh's tenure at Penn State, but he accomplished much in that time.

In Pugh, the bridge between the Euro-German university-science model and the American version and its culture is readily evident. He experienced the innovation of integrating teaching, graduate training, and abundant scientific research in German universities and became a full convert to the approach. In doing so, his intellectual stance rejected both the popular classicalist higher education curriculum in the U.S. and an emerging narrow vocational approach. Pugh's thinking represented a distinct third way of constructing universities by blending together the training of future farmers and agricultural scientists in good techniques and consumption of cutting-edge science that would be produced by the faculty themselves. This new pivotal cultural image was embodied by the faculty and students together plowing Penn State's land-grant fields, both to learn better farming practices and to do soil experimentation. The educational goal was not just to train farmers to be open to science, but also to train new chemists to do the science of soil for a better farm. The merging of education with scientific knowledge production was for Pugh, and others like him, the answer to ruinous fraudulent fertilizer scandals and other pressing issues facing American farmers. This American pragmatic take on the university-science model grew to represent an answer to what the environment, political machinations, and runaway capitalism could throw at the next generation of the nation's farmers. It also signaled a tilting of the qualities of knowledge towards epistemological privileging of systematic, codified, and universal knowledge promised by agricultural science over tradition, experience (sometimes even outright superstition), and a local knowing of the land.

Not surprisingly, Pugh's dream also included a pursuit of social progress, with a pious moral American Protestant bent. Pugh abhorred tobacco, alcohol, and promiscuity. Like other equally pious, Germany-bound American students, he must have made a strange sight among the cosmopolitan European students. Underneath many of these American graduate students' ascetic tastes, however, ran a deep tendency to think of blending advanced education, rational inquiry, and their uses in everyday life as an explicit path to achieving God's "city on the hill." While Christian ideology partially underpinned the rise of the education revolution in both the U.S. and Europe, religious motivation for a better society through education and science was an added stimulus behind the early Americanized university-science model and eventually did much to integrate it into the general culture.[8] And still today, the main ideology behind funding for science in the U.S., and increasingly elsewhere, is bound up in the cultural image of creating a better society, minus the overt religious trappings; a potent idea achieving far more mass appeal than an image of science as an independent discovery process for its own sake or only for narrow economic pursuits.

Also telling is that Pugh did not activate his dream from within the federal government. Without either federal ministries of education or science that could issue charters for universities or guide national education policy as in Germany at the time, this approach would have made little sense. Instead, Pugh had to start at the local level within an existing post-secondary organization. He did, however, famously work on the national scene as the only true scientist among the proponents campaigning for the important national *Morrill Land Grant College Act* of 1862. This legislation intended to reward states' loyalty in the brewing Civil War by a sell-off of federal land resources to start land-grant public universities, mostly in the mode of agriculture training and science (but it left future funding and operation to individual states).[9] While this did represent a kind of de facto chartering process, it had little future constraint on the eventual expansion of the scientific portfolio of these universities. By the twentieth century, as is still true, universities and local sponsors for the most part decided their own portfolio of graduate programs and academic departments, and they were unrestricted from competing for national prominence in research and other parts of the mission of a university. Also, the numerous stand-alone American undergraduate colleges could themselves choose to become universities at any point by simply adding graduate training. Naïvely, Pugh had initially envisioned establishing a centralized system of universities akin to his perception of the German system, but his subsequent local and national lobbying by writing newspaper articles, planning (he did author what became a popular plan for his vision of a university), and strategizing to win funding and public support for the ideas of science and its agricultural applications turned out to be exactly what had to be done in the decentralized educational environment of the U.S. At the same time, states had agricultural societies, essentially voluntary groups of wealthy farmers, who also advocated for inclusion of agriculture training and research in colleges and universities alongside other curricular pursuits. In this sense, significant parts of the Americanization of the university-science model were ground-up phenomena, literally born out of the soil and grassroots development.

Decentralization and profound localism, however, did not mean that Pugh's dream was uncontested. While specific historical details behind his struggles to achieve his vision are not essential here, suffice it to say that there were many, and his performance in overcoming them at a comparatively young age is a testament to his leadership. What is sociologically important, though, is that his struggle had to be simultaneously waged on two fronts, each representing in the second half of the nineteenth-century a strong cultural counterforce to the vision of a research university. At the same time, parts of each, mixed together, ended up supporting the Americanized university-science model, which is ironic given that neither's rhetoric seemingly yielded much space for compromise.

As alluded to earlier, on one side was a celebration of the practicality of farming through a deep cultural identification with the farmer, his family, and their trials in running the farm. This celebration became magnified in American society as change from industrialization and large-scale mercantilism threatened cherished myths of an orderly agrarian life. Nineteenth-century America considered itself a nation of farmers whose practical "know-how" was revered as key to the nation's success. Not surprisingly this know-how culture sometimes led to anti-intellectual sentiments. Hitching science to education, then, was no small feat in the decentralized U.S. When Pugh sought more advanced education for Pennsylvania's young farmers, he understood he was addressing a constituency from the nation's largest share of economic producers as well as its foremost cultural interest group. And as a result, his vision was not fully welcomed. Residing within this pro-farming culture was a suspicion of science, elite classical curricula, and all forms of advanced education. "Book farmers," the kind that Pugh wished to educate, were derided, while "practical farmers" were admired, as symbolized by the jokes about practical farmers' boasts of profits in front of the local university's agricultural experiment station. In many ways, this was the part of American culture that demonized the early stirrings of a broader education revolution—advanced education would supposedly lead rural youth away from farming and to the "temptations of an idle life" often equated with growing cities, a view that was the forerunner to current critics of the education revolution accompanied by their generally unsupported claims of overeducation, credential inflation, and superficial learning.[10] In the case of Penn State University, after Pugh's death the disappointingly modest amount of land-grant funds plus a new president who installed a more classical curriculum led the university away from Pugh's dream, almost to a premature demise. By the 1880s, however, the faculty had reorganized the university along Pugh's original vision, and from there it never looked back.

The second countering cultural force was a persistent notion of elitism from some individuals and universities in both education and science. If not quite as celebrated as farming, reserving advanced education at the university for an elite appealed as much as assuming science could be done only by the few most intelligent. Applying a "best and brightest" cultural notion to universities was also popular among scientists and educationalists at Pugh's time. A small, select group of prominent universities, such as Harvard and other members of the Ivy League, Johns Hopkins University, the University of Michigan (the first American university to employ the German seminar method of graduate study), the University of Chicago, and a few others, represented what, as will be shown further on, could have become a restricted set of universities more within the European mode. Supporting a culture of elitism was also the classical curriculum of literature, Latin, and the arts, then considered to be the pinnacle of advanced education and thus, the thinking went, appropriate for only a few. This course of study left little intellectual room for scientific research, and certainly no room for its practical applications. In addition, the initial take off of the education revolution was more about widening access to basic education than about access to advanced education for all, even though the latter eventually became a major outcome, as is evident in growing educational attainment in the country over the twentieth century.

While Pugh was comfortable with parts of the "best and brightest" culture, he certainly opposed the dominance of a classical curriculum as much as he championed science. His ideas, along with similar ones of many other educationalists, represented the beginnings of a third way, the long-term ramifications of which even he did not fully envision. Ultimately the vision grew and fed into a symbiosis among charters with flexible missions for universities, PhD training, and faculty research, all supported by access to growing advanced sectors of education within American society. This process produced an extensive supply of research universities and nearly inelastic demand for educational opportunity, ideologically driven by the education revolution. These struggling new universities, often miles away from any sizable city, with their "model farm" and its muddy research fields right on campus, were hardly recognizable as a budding Göttingen or Berlin, yet at their core an innovative version of the university-science model was beginning to unfold.

Even though Pugh's story demonstrates some of the larger cultural themes that went into fashioning the huge academic platform for STEM+ research, it is just one case. And the intention here is neither to Romanize it or make it more deterministic than it was. There were in fact scores of Evan Pughs, along with other founding organizations, including civic interests, religions, and a sundry of stakes in creating scores of universities that eventually became significant centers for STEM+, both public and private and combinations of both, across the country, each with their own historical twists and turns influenced by local champions and detractors. Observers of the capaciousness of the development of higher education in the U.S. always note this wide variety in the ways organizations are founded and what they include educationally at any one moment. Consequently, detailed histories must spend considerable effort in describing the bewildering number of paths taken in developing, changing, and often completely revamping their missions, particularly as they applied to research activities. By necessity it is a history mostly told one organization at a time. Rising slightly above all this variation, however, one can see that with little centralized control, the environment for founding, mission shifting, and development was comparatively open, and this is a key component of the access symbiosis. For example, concurrent with land-grant universities' efforts to merge practical knowledge with science training, there were diverse ways universities organized scientific and engineering research over the late nineteenth century. And Pugh's specific vision of agriculture and the university were continually merged into an explicitly academic model across many universities. The broader ideas that it tapped into, however, reflect how universities became so much a part of the culture and how this would eventually support many research-active universities in the coming era of mega-science.[11]

Beyond Wildest Dreams

Research-active, scientist-training universities are foundational to the capacity to do science. By 1920, the U.S. had thirty-nine such universities, then in 1950 it had more than tripled them, with capacity to train approximately thirty-five hundred new scientists annually. Building upon this, growth would continue to the degree that by 2010 there were over three hundred PhD-granting universities—all involved in research and training over twenty-five thousand new STEM+ PhD's annually. The consequences of this access symbiosis are illustrated in Figure 4.1.

The engine driving the process is represented by the line showing the steady increase in the proportion of American youth attending undergraduate, postsecondary education in all kinds of schools including universities. In 1920, the U.S. enrolled fewer than 600,000 students in all forms of postsecondary education, but by the end of World War II, the country enrolled approximately 2.3 million students per year. Undergraduate programs were never capped by any level of government, nor were university charters for merging undergraduate training with research activity and graduate training restricted. Thus, enrollment rates swelled from less than 20 percent of a youth cohort in 1950 to 40 percent by 1980. The shaded flows in the figure show growth in universities with varying levels of science production based on a currently used scale of research activity. The bottom darkest are universities that would come to have the highest research intensity, followed by medium-active and modestly active. Although producing less research than the most research-intensive, these latter two sets of universities also significantly contributed to the country's science capacity. As in Germany, early twentieth century expansion fueled the overall higher education system. But compared to Germany, the U.S. did this in a far more rapid fashion, providing the resources and human energy to culturally "cross-subsidize" science production in universities. This same process will be seen behind both the revitalization of post–World War II European science and the rise of scientifically productive countries in Asia and eventually elsewhere.

By 1900 a small group of elite, albeit not all private nor all old, universities were overwhelmingly responsible for the country's training in science and university-based scientific knowledge. What we can call here the "Elite-16," this group included the University of California at Berkeley, the University of Chicago, CalTech, Columbia University, Cornell University, Harvard University, the University of Illinois, Johns Hopkins University, Massachusetts Institute of Technology, the University of Michigan, the Uni-

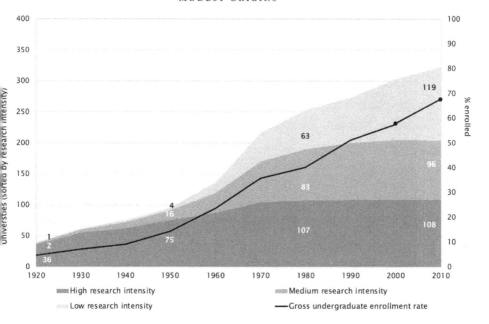

Figure 4.1. Consequences of the Access Symbiosis: Rising Undergraduate Enrollment Ratio and Growing Numbers of Universities, 1920–2010. Sources: U.S. Department of Education NCES: Digest of Education Statistics; U.S. National Science Foundation's Survey of Earned Doctorates.
Note: Undergraduate enrollment rate as proportion of all eighteen- to twenty-four-year-olds in any type of academic postsecondary school, according to data from the U.S. Department of Education. Number of research-active universities with STEM+ training calculated from NSF's Survey of Earned Doctorates. Carnegie research classification as of 2010: (1) universities with the highest research intensity (Carnegie R1); (2) universities with medium research intensity (Carnegie R2); and (3) universities with low research intensity (Carnegie R3). The classification was applied retrospectively, but verification of publications from selected universities show that later categorical membership approximates earlier production levels.

versity of Minnesota, the University of Pennsylvania, Princeton University, Stanford University, the University of Wisconsin, and Yale University.[12] These universities generated the bulk of university-based science from the middle of the nineteenth century until just before World War II, at which point they accounted for one-half of all university research expenditures in the country, according to the U.S. National Resources Committee in 1938. Historians rightfully note the unusual success of the Elite-16, and they could have easily become the foundation for an exclusive set of universities perpetuating the earlier European model. Consequently, these prominent universities and a few others like them are frequently credited with American STEM+ success, and their research capacity is certainly significant, but the larger story behind American science capacity is not bound up in just one small group of prestigious, wealthy institutions.

Indeed, the Elite-16 themselves foreshadowed the broader access symbiosis. Their founding and mission charters were neither issued nor controlled centrally, since many parts of the civil society could, and did, start universities. The Ivy League universities were mostly formed by various colonial Protestant denominations to train ministers and educationalists. Johns Hopkins University and the University of Chicago were started by educationalists and aging capitalists, or with their fortunes after their death, in an explicit attempt to emulate research activities at the nineteenth-century German research-intensive university. A bereaved wealthy merchant and his wife established the Leland Stanford Jr. University in memorial to their deceased son. The University of Michigan was founded by a mix of local, state, and religious authorities. Notably, the remaining public universities in this group were established through Pugh-like processes of blending federal land-granting and local interests in agricultural training and science. With flexible mission enactments, their future development and knowledge production focus was never limited, and most of them evolved in directions divergent from their original intentions. By the early twentieth century, too large a group to rely solely on training a small elite as their educational mission, they increasingly enrolled students from a growing variety of social and economic backgrounds.

Twenty universities were, by 1920, rapidly becoming research-intensive and would, with the original Elite-16, form the collective of the American Association of Universities (see darkest shaded base in Figure 4.1). Like earlier ones, these new arrivals on the scene came out of the same sundry of founding scenarios evident in the nineteenth century, and then, over time, entered into research and training of scientists. For example, The Ohio State University increased its training from finishing just two PhD students in 1920 to graduating more than fifty by 1930. Others such as Boston College, a Catholic university founded in 1863, and Michigan State University, a public land-grant founded in 1865, also did not start awarding STEM+ PhDs and undertaking serious scientific publishing until the late 1920s.

The way that the American system has allowed flexibility in mixing public and private funds at public universities is also part of the access symbiosis. Unlike in many other countries, U.S. public universities charge tuition (historically modest, now competitive with private levels) and have always been free to compete for funds from all types of sources, both public and private. For instance, between 1919 and 1925, five of the top eleven most successful national fundraising campaigns were conducted by state-public or land-grant universities. Over the first decades of the century, research-intensive universities grew at an exponential annual rate. At mid-century, science publication worldwide continued its exponential climb, and the supply of research-active universities in the country was growing by 3.3 percent annually, doubling about every two decades. And the most research-intensive universities would continue to multiply to over one hundred by 2010.

It is fair to say that as the twentieth century progressed, a degree of homogeneity emerged. Grafting research and STEM+ training onto universities, the country added innumerable other research-active universities, both as older, originally less research-oriented universities intensified their research mission (for example, the University of Texas, 1876) and newly founded ones included research as a mission from the start (such as the University of Houston, 1927). By the 1940s, a growing number of universities were following the most prominent ones to enact the university-science model and actively produce scientific research.[13] As a result, at present there are hundreds of universities that provide STEM+ PhD training and do extensive STEM+ research, about two-thirds of which are not of the highly prestigious, world-celebrated variety.[14] Keeping in mind that these groupings by research level are an approximation applied retrospectively, it does nevertheless illustrate how the process resulted in numerous universities involved in scientific training and research. The frequent enactment of the ideas embodied in Pugh's dream did not reproduce the most prestigious universities as much as it added a variety of universities doing research while increasingly educating all kinds of students.

Engine Within the Engine

The clear scientific success of the Americanization of the university-science model has captured the imaginations of late twentieth-century academics and science policymakers across the world, stoking their desire to emulate it in places far removed from its origins. But often these would-be imitators do not fully appreciate that the American extension of the European model is fueled by a culture of educational access, diversification, and a democratization of science. The world-class, often private, highly prestigious American research university charging expensive tuition fees and enjoying seemingly endless private funds indeed deserves much credit for what this country has contributed to mega-science over the century. Yet it is the aforementioned access symbiosis that is the real secret behind the success of the Americanized university-science model, as witnessed by the steady growth in large, mostly less prestigious public universities of all origins that Evan Pugh and others fought so hard for in the nineteenth century.[15] Over the years, large, accessible public universities, across varying levels of research intensiveness, have educated millions of undergraduates, and these universities' advanced science training forms the backbone of the STEM+ research capacity. Public universities are now responsible for about 60 percent of the country's annual new PhDs across all sciences and related fields, surpassing the private research universities in training females and individuals identifying with minoritized groups. Also, scientists at these public universities are annually involved in 45 percent of the country's STEM+ papers.

Although estimates suggest that scientists at the wealthiest, high-prestige, private universities publish a modest amount more per researcher than those working at most public universities, the scale of the latter is formidable. For example, consider an organizational representation of the university-science model by the College of Science (known as a Faculty in most other countries) at Evan Pugh's land-grant Penn State University in 2016–17 (Figure 4.2). At the top of the pyramid are 421 permanent faculty-scientists, who along with 934 new PhDs (postdoctoral researchers) and graduate students undertook research that year with approximately $100 million in research expenditures from a mix of competitively gained government

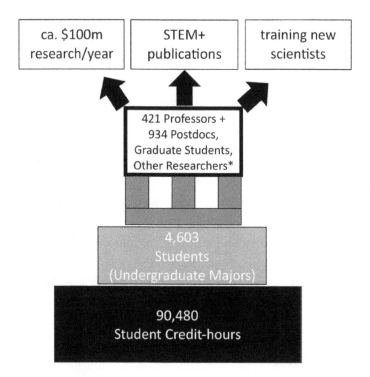

Figure 4.2. Implementation of the University-Science Model in a Large College (Faculty) of Science in a Public U.S. University. Source: Data from College of Science, Penn State University, University Park Campus, AY 2016–17. Note: Enrollment expressed in student-credit hours.
*Also taught by 110 "teaching only" instructors.

and private external funding. They are supported—cross-subsidized—by over 4,600 undergraduates majoring in natural sciences, and along with nonmajors contribute to an annual 90,480 student credit hours in the college's courses. While the faculty-scientists do teach, instructional loads take up only a small portion of their time, and it is common for them to spend as little as six hours a week in class and the rest on research. The College of Science also relies on about one hundred full-time instructors who are generally not involved in research. In addition, this is only one of eight similarly structured colleges, at this single university, all contributing to the university's annual total of approximately three thousand (in 2011)

STEM+ papers, with most research-active faculty publishing an average of four to five coauthored papers annually, and sometimes many more.

This access symbiosis and its influence on expanding the university-science model has played out many times over. Consider one last example. Texas Tech University, in a small West Texas town, began as a modest-sized technical college in 1923 with a narrow range of training and not much of a research profile. In the relatively open educational environment in the U.S. since then, it steadily added various parts of a complex university, and now enrolls forty-thousand students with a full curricular profile and faculty undertaking research. Employing the exact same organizational strategy shown earlier across its full set of colleges in STEM+ areas, the university now produces significant research output and in 2016 was classified among the country's high-intensity research universities (R1). It is this process and the Texas Techs, as much as the Harvards, Michigans, and MITs, that provide the vast platform for the American contribution to mega-science.

CHAPTER 5

C. N. Yang's Children

Globalize or Fade

During the 1970s, just underneath the humming U.S. science production, loomed a crisis: the number of new scientists had hit a plateau that, if continued, would greatly curtail the country's contribution to the growing globalization of science. This was a significant irony given all that the Americanized university-science model had done for the pace of new scientific discovery up to that point. The explosion of research universities described in the previous chapter made for a dramatic expansion in the supply of graduate training programs in science. Not every new STEM+ PhD goes on to publish new science, but a significant majority do, and a continual refreshing of the pool with young scientists is essential for the capacity to do scientific research. Over the early twentieth century, this feat was almost completely accomplished by enrolling white, American-born males into graduate school (as shown in the darkest shaded stream at the bottom of Figure 5.1). Accompanied by a small group of white females, 86 percent of the new PhDs in 1920 were American white males, and by 1950 this group's proportion had increased to 95 percent of the thirty-three hundred new scientists that year, in part motivated by a large U.S. government program to increase postsecondary education. But in just three decades this situation ended. By 1980, the volume of new STEM+ PhDs from this demographic group had fallen back to the lower level of 1970. Setting aside the 1965 to 1975 spike from the baby-boom bulge and those men seeking graduate educational deferments from the Vietnam War, if PhD training had stayed limited mostly to white American-born males, the country would not have had enough new scientists to remain central to a maturing global mega-science. The number of white American males flatlined after 1980 and proportionally dropped to 28 percent of new STEM+ PhDs by 2010. Given the already large growth from this group, even if efforts were made to recruit more of this group into science, it is doubtful that there would have been enough

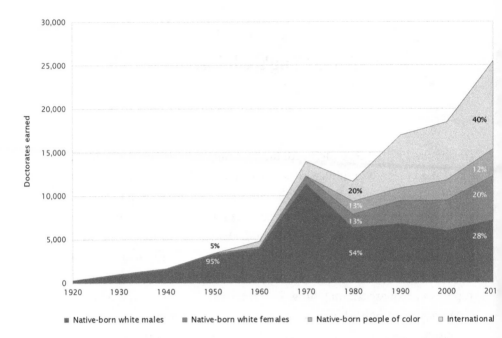

Figure 5.1. Diversifying Flow of STEM+ PhD Graduates from U.S. Universities, 1920–2010. Sources: U.S. Department of Education NCES: Digest of Education Statistics; U.S. National Science Foundation's Survey of Earned Doctorates.

new enrollees with the required aptitude and motivation to significantly change the situation.

What happened next is clear. The access symbiosis that grew research universities also led to a widening diversification of students pursuing advanced degrees in science at American universities. Since the 1970s, a persistent, albeit partial, diversification of science training saved the science capacity of the country. A combination of white American females and people of color from marginalized groups increasingly entered STEM+ graduate programs. By 2010, white American-born females had grown to be a fifth of all new scientists. The smaller proportion of people of color among new PhDs into the twenty-first century remained constant since 1980, but their absolute numbers modestly grew over the next thirty years with the overall expansion of STEM+ training. Still well below their overall share of the country's population, these groups nevertheless elevated research capacity. And inspired by the culture of the education revolution, the goal to generate even more inclusion in advanced STEM+ education from less-represented groups is a constant focus of federal and local university programs. A second substantial source of this were the numbers of international students, mostly with undergraduate degrees from outside the U.S., which by 1980 represented one-fifth of new PhDs across all the STEM+ fields, and by 2010 had grown to 40 percent of the total pool of twenty-five thousand new scientists.

While the general story about international students is known in U.S. science policy circles, both its origins and full implications about the role of educational developments in science are not. Universities, of course, were born out of globalization, or at least the pan-European version of globalization some eight hundred years ago. They have always been open, by varying degrees, to students from beyond their local region; traveling scholars were given local rights based on charters of universities that transcended the authority of territorial princes, bishops, municipalities, and other local authorities. This centuries-old tradition foreshadowed a trend that would pick up speed and gain normalcy. Certainly, early American universities had the occasional international student and, as already noted, the German innovation of the research university attracted talented students from across Europe, North America, and beyond. But in the latter half of the twentieth century the American research university would attract even more international students. Much like the growth in the numbers of public and private research universities, this process was not a result of a national or any local state's master plan, or even much of an explicit strategy by specific universities.[1] Instead, it stemmed from the improbable intersection of the cultural forces behind the expansion of higher education and the country's foray into an imperialist resource-grabbing frenzy in Asia, fueled by the ugly side of late-nineteenth-century American hubris.

C. N. Yang: Epitome of Human Capital Flow to Universities

In the summer of 1900, persistent attempts to carve up China for commercial exploitation and Christian conversion by six European powers, along with Japan and the U.S., precipitated the horrific, antiforeigner Boxer Rebellion and the equally bloody gunboat response from the colonial powers. With considerable looting and atrocities along the way, upon the eight-nation expeditionary force's military occupation of Beijing, China was forced to pay over the equivalent today of US$60 billion, with shares going to each invading country as "reparations" so that they would vacate the country. Later, when it became clear that the U.S. had mistakenly received a considerably larger payment than initially agreed upon, a political debate ensued over how to use these funds, and although the Chinese royal government was desirous of the funds' return, the American administration never seriously considered repayment. Instead, in 1906, the U.S. decided to use the overpayment to finance a scholarship program, sending Chinese students to study at American universities and return them home again as Western-style leaders. The overarching idea was that for China to join modern nations and be a better trading partner, it had to abandon its traditional elite, replacing them with Western-educated, technically knowledgeable leading citizens to guide it. And the way to produce these, it was assumed, was to expose selected students to higher education in the U.S.

While an unlikely outcome from a military conflict and clearly one with much hegemony foisted on China, nevertheless one can see in the "Boxer Indemnity Scholarship Program" the same blend of American cultural forces responsible for the rise of research universities, which were often the ultimate beneficiaries of the program. Commercial interests were certainly in play, but so was the notion that university training is more than classical training of elites and could be interjected more broadly across any society. The thinking was, if the university can make better citizens for the U.S., why not for China too? And last, since many Protestant missionaries to China had colleagues slaughtered in the initial uprising, the idea of advanced education as a gateway to a Chinese Christian "city on the hill" was also appealing in the early twentieth-century U.S. As the cultural matrix of the American education revolution heated up, it is not surprising that Edmund James, with a PhD from Germany's University of Halle and the president of the University of Illinois, an early productive public research university and one of the "Elite-16," was an initial proponent of this education solution. He and other leaders in higher education helped convince President Theodore Roosevelt to establish the scholarship program.[2]

After the collapse of the Qing dynasty, China turned neither to Christianity nor to Western liberal society. And although some American-educated Chinese returned as political advisors to various competing warlords and the early republic before the rise of the communist regime in 1949, the scholarship program never came close to its original goals. Unexpectedly, however, it did help to regularize the flow of international students to American universities, fostering a new stream of scientific talent to be incorporated into the Americanized university-science model.

Born two decades after the Boxer Rebellion in the then-quiet market town of Hefei, China, Yang Chen Ning, or C. N. Yang, as he has come to be known worldwide, embodies a process that eventually led to international students being a large part of science capacity in the U.S. Gentlemanly, inquisitive, and open, he was ever the consummate colleague at the institutions where he researched. Awarded the Nobel Prize in 1957 for the "Yang-Mills theory," established through papers on particle physics he wrote while on faculty at Princeton University, Yang is exceptional in many respects. Yet his story typifies the role of the access symbiosis in boosting the science capacity of American universities just in time for the coming intense globalization of science.

The internationalization of American STEM+ training has several important components. First, advanced students with considerable previous training in science came to the U.S. to pursue the PhD and subsequently took postdoctoral positions that centered around intensive research activity. In Yang's case, before enrolling for the PhD at the University of Chicago in 1946 and undertaking a postdoctoral position in 1949 at the Institute for Advanced Study at Princeton, he had already earned the BA and written a scientifically sophisticated master's thesis in China. Given the relatively rare talent and extensive prior training it takes to contribute to theoretical physics, students like Yang represented a major advantage for the American university. An increasing international graduate student flow meant that American universities' pools of prospective qualified individuals also increased.

Second, once these international students were trained, the expanding STEM+ capacity of the U.S. was able to absorb them, and this was particularly notable from about 1970 onwards. The access symbiosis—linking expanding demand for postsecondary training with knowledge production capacity—meant that newly trained PhDs had reasonable chances of becoming faculty. Yang, for example, was recruited by the State University of New York at Stony Brook in part to build research capacity in physics. Although there were periods in the 1980s of tight PhD labor markets, in general the incentives of universities to train international students occurred in concert with the prodigious growth in American research universities' dependence on a steady pool of young faculty-scientists.

Third, although this is now changing, international students are not always interested in returning home to pursue careers, especially due to limited opportunities for research. Although Yang became the first highly acclaimed, U.S.-based Chinese scientist to visit the People's Republic of China in the 1970s and subsequently assisted in the opening up of Sino-American relations, he would likely not have had much of a scientific career through the political and cultural turmoil over the middle decades of the century in China. Even though after the 1950s the Chinese regime disallowed scientists to train in the U.S. until 1980, aspiring researchers from many other countries flowed into American universities. As the percentage of international students earning PhDs in STEM+ fields increased, so too did the percentage who wished to stay and work in the U.S. In 1995, one in two international students who completed their PhDs intended to stay in the U.S., and by 2015, approximately 75 percent of international PhD earners intended to pursue a research career in the U.S.[3]

C. N. Yang was certainly not the U.S.'s first Chinese international student. Indeed, his father, K. C. Yang (Wu Zhi), obtained a PhD in mathematics also from the University of Chicago some years before and then returned to China. And, of course, most scientists, international or not, do not win the Nobel Prize, nor do all international students end up as faculty at American universities. But just past mid-century, Yang's story foreshadowed what would become a major trend in a flow of talent into universities, which by the 1990s would be an essential component of American science capacity. Plotted against the increasing volume of international students earning STEM+ PhDs since 1960, Figure 5.2 shows the proportions from the three largest sending countries of China (plus Taiwan), South Korea, and India, and all other countries. At the beginning of the growth in the trend, most students came from other countries, but as international STEM+ students became a significant flow by 1980, students from China and Taiwan greatly increased. By 2010, Chinese students, along with students from India and South Korea, made up approximately one-half of this trend. Consistent with the access symbiosis argument, students from these three countries have also made up the highest proportion of international undergraduates in recent years.

But the trend in graduate study is not only related to a few Asian countries—it always was global and has gotten more so over time. Over the past fifty years, in addition to the three large Asian senders, PhD STEM+ students have come from a diverse set of countries from all regions of the world, including Brazil, Canada, Egypt, Germany, Greece, Iran, Mexico, Nigeria, Thailand, and Turkey. Of course, geopolitical machinations can influence shifts in origins, but the remarkable openness and autonomy to participate in this trend finds American universities also training students from many decidedly non-allied countries. Among all other sending countries, for example, the fourth-largest number of students came from Russia in 2005, and Iran has consistently been among the top ten senders before and after its fervently anti-American revolution, coming in third in 2010. And, as would be expected given the access of American universities for developing their own science capacity, this trend was not limited to the most prominent universities and is currently widespread across a range of research intensity, as well as both public and private control. By 1960, one hundred universities producing most of the STEM+ publications had international STEM+ PhD students, and over the next fifty years this would triple to approximately three hundred universities, including the most research-active universities in the country.

This story of C. N. Yang is one of extraordinary success. Although he exemplifies the best outcome of this trend made possible by the American university-science model, there can also be a dark side to the access symbiosis. While these students' essential contribution is recognized, they have been stigmatized at times as presenting significant challenges to university life, and universities have often lagged in providing services to their growing populations of international students.[4] Much like the enduring paradoxical celebration and condemnation of immigration throughout U.S.

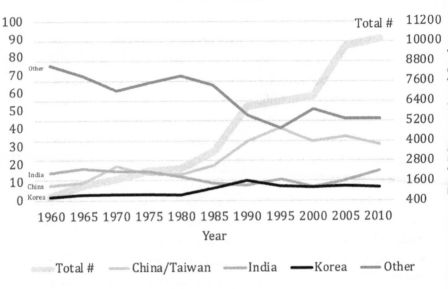

Figure 5.2. Awarded STEM+ PhDs to International Students at U.S. Universities, 1960–2010. Source: U.S. National Science Foundation's Survey of Earned Doctorates. Note: Until the 1980s, the flow of Chinese students was entirely from Taiwan.

history, the internationalization of advanced science training confronts a similar paradox within the broader American culture.

UNFOLDING ACCESS SYMBIOSIS

Whether judged as an access glass half-empty or half-full, a maturing of the education revolution continues to be a key component for this country's ability to be at the center of mega-science through to the present. The unprecedented mid-century increase in secondary education and then undergraduate enrollments, which Parsons and others observed as so transforming, swept into graduate training at more universities for students with a broader array of backgrounds. Figure 5.3 displays some notable milestones behind this success in educating more scientists and expanding research capacity. Consider gender diversity. In 1950, for example, 5 percent of new PhDs were U.S.-born women, a share that would double by 1976, and double again to 20 percent in 1980, remaining at that level, albeit of an expanding total, through 2009. At the same time, by 1966, female students accounted for 40 percent of all undergraduate enrollments.

This percentage continued to increase over the remainder of the century, and there are now more female than male undergraduates. Similarly, in 1945 public universities graduated half of the total supply of American undergraduates who later earned STEM+ doctorates, and by 1990, more than two-thirds of first-generation (that is, first in family to attend postsecondary education) undergraduates who went on to be PhD scientists originated from public universities. And while access has moved more slowly for people of color, there has been some growth. For example, fewer than 50 percent of Black high school graduates enrolled as undergraduates in higher education in 1972, but this percentage grew to almost 75 percent by 2004. And, between 1974 and 1995, the percentage of STEM+ doctorates earned by U.S.-born people of color increased from 13 percent to 18 percent, yet this share decreased by 2010 as the relative share of STEM+ PhDs increased more quickly among international students. And this inclusion has not been even across all universities.[5]

SCIENTIZING THE REST

The last story of the American experience with the university-science model illustrates how far the relatively open charters of the country's postsecondary institutions have led them into scientific research. The earlier trend of ever more research universities joining into the country's science capacity has since about 1980 spilled over to all types of colleges and universities. The entire spectrum of postsecondary institutions is now involved in research. Not only do STEM+ papers routinely come from scientists at smaller, mostly master's-degree-granting universities historically not involved in research. As shown in Figure 5.4, faculty-scientists from the traditionally nonresearch colleges, small universities, and community colleges were involved in about eighty-thousand papers in 2011, or nearly a third of all papers with at least one U.S.-based author. And these schools are involved in all dimensions of mega-science, including external funding, international collaborations, research done on a variety of subtopics of science, and a steady increase in the average quality level of papers (measured by journal impact factor). The highest volume among traditionally nonresearch postsecondary institutions is from the smaller, often master's degrees only, universities, but all types have increased scientific

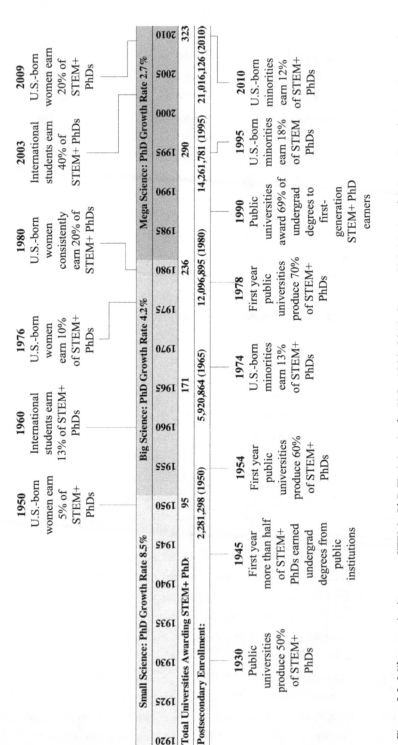

Figure 5.3. Milestones in Access to STEM+ PhD Training in the U.S., 1930–2010. Source: U.S. National Science Foundation's Survey of Earned Doctorates. Note: The terms small, big, and mega-science here are used for better visualization of the historical trends, but these should not be taken as a sharp demarcation of periods.

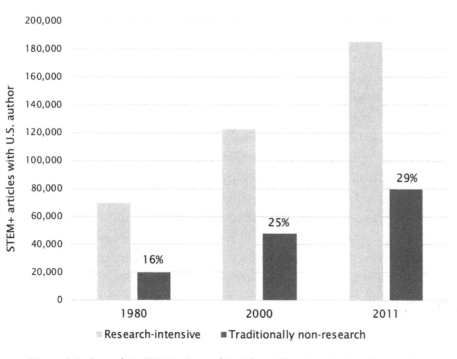

Figure 5.4. Growth in STEM+ Journal Articles with a Scientist-Faculty Author from Traditionally Nonresearch Postsecondary Schools, U.S. Source: SPHERE database. Note: Traditionally nonresearch includes institutions categorized by the U.S. Carnegie Classification as doctoral nonresearch, master, baccalaureate, associate (community colleges), and professional schools and seminaries.

output by four to six times over the past several decades. The number of traditionally nonresearch postsecondary institutions regularly involved in research more than doubled in thirty years from just over three hundred in 1980 to approximately seven hundred by 2010. Perhaps most illustrative of this trend is that more of the approximately fourteen hundred associate's-degree-granting, two-year colleges across the country are also contributing STEM+ research.

Professor Shannon Smythe Fleishman's open face beams with a knowing smile when she talks about the future of two-year colleges, referred to often as community colleges. These schools have a long history in the U.S., most often associated with providing a range of educational programing for local communities in gaining required skills beyond secondary schooling for jobs in health, social services, business, and others; assisting youth and adults alike to recover from unfinished or lackluster high school experiences; and providing a stepping-stone to four-year colleges and universities. Fleishman, as Department Chair of Social Sciences at Chesapeake College, a smaller than average community college just across the Chesapeake Bay from Annapolis serving five rural counties of Maryland's Eastern Shore, has observed the university-science model slowly but surely making its way into these colleges. She describes how, as part of what has been an intensifying trend over the past several decades, her campus has increasingly engaged in STEM+ research. One of her colleagues, for example, collaborates on analyses of data from the Human Genome Project, while another is involved in research on renewable energy and environmental sustainability projects, both with the usual grant support and, of course, leading to paper publications. These are, as Fleishman describes, "passion projects" for faculty, not necessarily required for promotion, and partially dependent on collaboration with other researchers at larger, more scientifically equipped universities. Nevertheless, research projects like these are becoming ever more common at community colleges for numerous reasons.

In the past, most faculty at these colleges did not hold the PhD, but that is changing rapidly. As Fleishman points out, most of her local and national peers have been extensively trained in PhD programs at universities, so they come with "a belief in science, the necessary technical skills, along with a passion for the craft of research of all types." Once exposed to the university-science model with its image of faculty as teachers and knowledge producers, it is hard to resist. And while the average teaching requirement of five courses per semester is over double that of typical research universities, some community colleges are leaning towards reducing faculty teaching load in exchange for time to undertake scholarship. For example, the seven community colleges of the City University of New York, one of the largest urban systems, has lowered teaching loads to make more time for scholarship, including scientific research, and is requiring research publications for promotion of faculty.

Also, the education revolution's ideas of an expanding postsecondary experience for more people are at work here too. With approximately one-half of all undergraduates in the U.S. entering a community college, these institutions include students from a range of backgrounds with a wider set of educational goals than was historically the case. As Fleishman notes, the higher-performing, aspiring youth from more educated middle-class families are increasingly starting their college careers at less expensive, closer-to-home, two-year colleges with the plan to eventually move on to universities. What was once engineered as a first phase of postsecondary education in only some states, such as California, is now spreading, and with it come incentives for community colleges to expand their academic offerings. Along with targeted programs such as academic honors, curricular tracks, and competitive scholarships with links to acceptance as a midstream undergraduate at states' public universities, faculty increasingly see inclusion of more students in research as a valued educational objective. And then as more students with an aptitude for research are motivated to participate as productive research assistants, a reinforcing process leads to more science from this sector. It is not surprising that about a fifth of annual new STEM+ PhDs started their undergraduate training at a community college, representing a full third of U.S.-born PhDs.

These factors continue to scientize this large sector of postsecondary education, along with the newer inclusion of all kinds of universities and colleges not traditionally involved in high-level science. What started as a modest idea, chaotic and somewhat unintentionally in the U.S., set the European successful university-science model on a new course, one that eventually opened to a logic that advanced formal education could be, and should be, about the human development of many individuals, not just a small elite, and about achieving the good society. And with that idea came great support for the activities of science and research. At the same time, with this new model of advanced education and the symbiosis of access to new students, new research universities followed by all other postsecondary institutions drove the acceptance of greater participation in higher education deeper into American culture. This melded expanding higher education with scientific research into a cultural form that was no longer considered as elitist, restricted, or necessarily disconnected from everyday life. Even though American society is, in the abstract, no more or less scientifically astute or supportive of science than any other educated society, the country's capacity to undertake huge volumes of science is in part due to its connection to a celebrated and growing collegiate culture. But this is not often recognized. It is a common mistake to assume that the dizzying array of nonacademic activities embraced, administered, and funded by the average American university, from a near fetish over collegiate sports to religious services at public institutions to sprawling "student life" complexes, are mostly to legitimate, somewhat cynically, the institution within the culture. Universities are not cherished so much, for example, because of their sports; their sports are cherished because they are part of the highly valued, widely accessible collegiate culture. They are themselves a result of the path that Evan Pugh and others like him set towards making American higher education an essential, almost everyday, component of society. All these nonacademic activities grew out of the considerable legitimation that this new university model eventually achieved in mainstream American culture, and science and research went along with it. Even the extensively noted excesses and inequalities of higher education are not so much an attack from the outside on the academic integrity of the university as they are a product of the relentless binding of the university with everyday culture. Hence, they are seen as shortfalls in achieving that valued goal.

Certainly, there are elite and extremely expensive parts of this culture that are not accessible to the many, but the vast majority of postsecondary schools in the country are at least modestly accessible, and they are now all producing research with some version of the university-science model as a guide. How sustainable that model will be into the future will be explored in the last chapter, but up to this point in the historical journey, it is a robust cultural engine. And this is what Parsons's sociological insights were onto: the university's charter and ability to generate new knowledge, science included, had in the American case become turbo-charged by the ideas of the education revolution that continued to increase the intensity of the relationship between the university and society rapidly and profoundly. From the unfathomably resourced, world-renowned super research universities to the modestly funded, often overlooked community colleges, the university-science model saturates the culture, scientizing ever more faculty, students, and organizations of advanced education. But is this highly effective university-science model the only one that was possible, or has it out-competed other models for supporting mega-science? Returning to Germany, we discuss this counterfactual argument next.

CHAPTER 6

The Theologian's Institutes

A Culture of Scientific Genius as Counterfactual

One way to judge the centrality of educational development and the Humboldtian perspective in the university-science model is by considering a counterfactual process. In other words, could global mega-science have been created by a different model, one not directly connected to the university and the education revolution? If so, then our argument is weakened. With some irony, given its leading role in the origins of the research-intensive university, Germany provides an instructive counterfactual: the independent, government-funded, and highly prestigious research institute, autonomous from universities, fully dedicated to scientific discoveries, and often explicitly built to feature selected "genius" senior scientists. As of 2010, Germany was the world's third-largest producer of scientific research, achieved in part by its large array of research institutes, supported by a "dual pillar" science policy that purposely diminishes the full incorporation of the university-science model. The degree to which universities and the university-science model fared under this policy is a useful indicator of the robustness of our argument. To foreshadow the answer, the university-science model remains robust despite a competing, and well-funded institute model operating in parallel. This conclusion will bring us to an analysis of the broader scientization of all kinds of research organizations, also built upon the fundamental scientific platform of universities.

THE HARNACK PRINCIPLE

Through a chilly rain on a late October day in 1912, Kaiser Wilhelm II, uniformed and helmeted in the distinctive Prussian *Pickelhaube* with saber at his side, paraded with his honor guard into the new, well-appointed Kaiser Wilhelm Institute for Chemistry in Berlin's leafy suburb of Dahlem. Commemorating the first of what would be a group of institutes each dedicated to a different scientific field under the umbrella of the Kaiser Wilhelm Gesellschaft, or Society, the emperor was feted with flashy chemical demonstrations including close exposure to three hundred milligrams of a radioactive substance whose danger was underappreciated at the time. The kaiser endured the radioactive tour because, even though German universities were producing the most science in the world, his advisors were of the strong opinion that more scientific advances were needed to ensure Germany's new economic and military world prestige. At this time, science was increasingly seen as part of statecraft to be directly fostered by the government. So the parallel development of the society's institutes was to be for a qualitatively better, more intense, scientific environment compared to that found in universities.

The choice to direct the establishment of the Kaiser Wilhelm Society was Adolf von Harnack, a renowned professor of theology at the University of Berlin, whose academic career would culminate in some sixteen hundred publications on the intricacies of the history of Christianity. A slight, bespectacled man with a high forehead and fine features, Harnack looked every inch the scholar engrossed in a university's musty archives of antiquity. Indeed, he was called to a professorship at the University of Berlin, where he wrote widely discussed, controversial books employing a new historical-critical method. Essentially, he argued that the New Testament Gospels carried the essence of the religion, while various ecclesiastic arrangements institutionally necessary for Christianity's survival over the past two millennia, including Luther's reformation, resulted in spiritually limited, unnecessarily authoritative dogmas. While it was at the cutting edge of prestigious historical inquiry taking place at European universities at the time, Harnack's thesis did not earn him favor with the state Lutheran church, whose administrative hierarchy unsuccessfully tried to block his faculty appointment in Berlin. Although the contemporary field of history is considered empirical but not a science, since in Germany at the turn of the twentieth century this academic topic was newly empirical (using historical record as verifiable evidence), it was assumed to be part of the broad activity of *Wissenschaft*, the production of new knowledge. Those who undertook it therefore carried considerable cultural prestige, hence the attention given to Harnack's appointment by the Lutheran elite.

Yet, even as a "scientist" of church history, a theologian would seem an unlikely choice to lead a complex new association of specialized research institutes. Harnack, though, had additional notable qualities.[1] Elected to the Prussian Academy of Science, raised by the kaiser to the honor of hereditary nobility of the Empire, and named a high-level advisor to the Ministry of Education, Harnack was connected to the monarchy and its administration. He also brought to the position an uncanny ability to lead complex organizations, as demonstrated while he was head of the Royal Library in Berlin.

The initial idea was that the institutes would be connected to industry to maximize the application of new science to economic benefit; thus many industrialists were on the original board of directors, leaving Harnack the only academic among them. The eventual twenty-one research institutes of the Kaiser Wilhelm Society were initially resourced with private funds from industrialists, but with considerable legitimation and prestige granted directly from the Crown. University-industry research collaborations had already occurred, such as Justus von Liebig's 1840 research laboratory at the University of Gießen, and later Carl Duisberg's scientific leadership at Bayer AG. But the society's institutes came to represent much more than R&D collaborations with business as they steadily moved towards basic independent research. And the brilliant Harnack, although he would not be considered a scientist today, embodied the right cultural spirit and knack for administration to head the society. He did so right up until his death in 1930 at the age of seventy-nine while traveling to Heidelberg to start yet another new institute, this one focused on medical research.

If, as illustrated by Göttingen's development, the Humboldtian research ethos permeated the development of the research university, then the related ethos of the "genius" infused the development of the institute. And while the implementation of each ideal over the twentieth century led to a different path for the development of research capacity, they stem, and continue to receive significant legitimation, from a former shared cultural idea. Like the holder of a chair-professorship (*Lehrstuhl*) once representing the pinnacle of intellectual preeminence for an entire academic field in the German and European university, a central motivation behind the organization of the institute was, and still is, to maximize the contributions of an identified scientific genius, ideally in the prime of their career, sitting at its head as director. More valued even than maintaining a rationalized topical differentiation of institutes across the sciences, this "built around a genius" approach initiated, funded, and dictated closure of specific institutes upon the retirement of the genius at the state required age of sixty-five. The genius ethos, and the selection process, have become known as the "Harnack principle," and the institutes that it produced came to be at the core of Wilhelmine *Sammlungspolitik*, the politics of national cohesion for the new empire, formed to concentrate within organizations the intellectual achievements of the German nation.[2]

After the fall of the empire in World War I, these institutes carried on, but later disastrously colluded with the National Socialist regime, including criminally providing technical assistance for the mass murders of the Holocaust and other atrocities. Like other German institutions corrupted by fascism, the society was disbanded after World War II and its administrative leaders tried and imprisoned—hardly the future envisioned by their founding genius theologian. In 1948, following the postwar political partition of the country, with safeguards of clearer legal autonomy from political involvement but otherwise still based on the Harnack principle, the government of West Germany reestablished new independent research institutes under the umbrella organization of the Max Planck Gesellschaft, or the MPG as it is known for short, named after the founder of quantum theory who was also a past president of the Kaiser Wilhelm Society. And it is with the MPG that the counterfactual analysis begins.

THE DUAL PILLARS

When entering the MPG headquarters in Munich, just after passing by a statue of Minerva (the Greek goddess of wisdom, who serves as the society's logo) one must walk over a thin moat-like flow of water on a stone footbridge. Upon this bridge is inscribed in ancient Greek, "Full of desire and love, always to be the best and distinguished above the sum of all others." Unabashedly, the Harnack principle lives on. A contemporary report describes the scientific conditions at institutes of the MPG: "The scientific attractiveness . . . of the Max Planck Institutes are built up solely around the world's leading researchers, who themselves define their research priorities and are given the best working conditions."[3] A director not only has full power over hiring scientific staff and allocating research budget, but he, or only rarely she, has full authority over the course of scientific research of the institute (or a full department of a larger institute).[4] Initially, the institute form was not meant to be a wholly different organization for doing science and in many ways was an extension of one part of the original design of the university: the professor as an autonomous generator of new knowledge. The Harnack principle, along with significant resources, lives on and takes this a step further. From striking, architecturally unique buildings to self-celebration of its ethos, the institutes capture considerable scientific resources and prestige, thus perpetuating the logic of what is known as the *dual-pillar* science policy.

Since the 1950s, the guiding principle behind research policy in West Germany and then carried over after reunification is the image of universities and institutes as supporting but differentiated pillars of science production. While universities undertake research, they are assumed to specialize in training new scientists. The renowned and well-resourced research institutes do less training than research and are assumed to produce most of the country's STEM+ research. Conversely, the institutes' role in scientific discovery enjoys a kind of "favored sponsorship" by the state in terms of resources, which is reflected in enormous prestige within the general society.[5] They are considered organizations of the "best and brightest" researchers guided by prestigious senior scientists in the systematic pursuit of the most cutting-edge science possible. Their overall conditions for research are characterized through well-resourced, dedicated—and interlocking—streams of federal and state (*Bundesländer*) funding, albeit recently with some competitive (peer-reviewed) revenue streams added in and becoming increasingly important. While the early charter of some institutes involved partnerships with industry, they were never limited to research primarily for economic, military, or state purposes. And since their origins, they have evolved into centers of fundamental scientific inquiry to be shared openly with the world's scientific community in the standard fashion of journal articles, conferences, and so forth.

The fate of the research university in Germany has been much the opposite under the dual-pillar policy. Since the 1960s, massive tertiary educational expansion did revive research, but the dual-pillar policy established a pattern of chronic underfunding, in contrast to the enrollment-induced equivalent investments that have expanded research capacity in so many other countries.[6] By 2017, even though Germany spent the most among all European countries on R&D (3 percent) relative to its high GDP, after industry's share, its universities received only 17 percent of these funds, while a significantly larger share went to support research in the well-resourced institutes. Because unexpectedly rising student enrollments were not accompanied by proportional increases in hiring of professors or senior research staff, German universities' larger teaching loads reduced their members' time for research. Since World War II there has been only a partial enactment of the cross-subsidizing of the university-science model compared to the average research-intensive American university. Undertaking research still is part of the German university's mission, of course, but this role was assumed to be steadily—and properly—eclipsed by the institutes. Therefore, most university-based research must rely on competitive R&D funding, while institutes enjoy permanent dedicated funding, cutting-edge infrastructure, and far more time dedicated to research.

And this has continued as mega-science has heated up. By the 1950s, the *Bundesländer* chartered, funded, and managed universities, trapping them into a zero-sum dilemma between using set funding allotments for both teaching and research expenditures; while pressure from the former rose with more enrollments, the latter suffered. Instead of resource increases equal to the massification of tertiary education, research universities have steadily been taxed with educational tasks such as training approximately two-thirds of the country's total postsecondary enrollment and generating a high PhD graduation rate, both among the highest rates of all developed national postsecondary systems. Further disintegration of teaching and research also occurred through the establishment of non-PhD-granting universities of applied sciences (*Fachhochschulen*), devoted more to teaching in technical fields than research. Although gradually the *Fachhochschulen* have increased participation in research, it is mostly in collaboration with industry on applied problems; only rarely are they able to confer doctorates.

All these factors are responsible for a long-simmering national "crisis of legitimation" of German universities and their role in science. The steadily rising enrollment, heavy teaching loads, and steadily increasing competition for a share of research funds leaves universities having to do more with less. Universities, many of which have played crucial roles in early mega-science, neither want to do less research or accept a dual-pillar identity. Flying in the face of the policy, reflected in German media's frequent message, is a national aspiration to have its universities be world class again. And yet only a few are considered "world leading" in widely publicized international rankings, despite multiple cycles of an "Excellence Initiative" policy designed to select and valorize some leading universities thought able to compete globally at the highest level. The same issue can be found in France, where a recent research policy addresses a perceived problematic gap in both funding levels and prestige favoring research institutes, especially the CNRS laboratories, while universities are still viewed primarily as teaching oriented; the latter are being merged regionally, ostensibly to achieve critical mass to better compete globally, underscoring the power of comparative performance measures and rankings.[7] An accentuating sense of crisis among universities with a concurrent celebration of the autonomous research institute promotes the public perception that Germany's prestigious research institutes are where almost all its significant science is conducted, and that many universities are receding in research and are mostly training new scientists.[8]

Production Across the Sectors

But this perception turns out to be a myth. It is true that the 427 independent, nonuniversity institutes that produce STEM+ research, coordinated under the MPG and within three other umbrella organizations (Fraunhofer Society, Helmholtz Association, Leibniz Association; each with a unique profile of big-science research facility management, applied contract research, or basic research, respectively), are indeed very successful in tackling important scientific questions. Often, they serve as catalysts in highly specialized subfields, relying on unique conditions and infrastructures; however, this leads to a research system that is highly fragmented institutionally and hinders effective knowledge flows.[9] Despite less research funding, less-than-optimal research environments, and the added responsibilities of teaching and training, the country's 142 research universities produced the largest share of the country's contribution to worldwide STEM+ papers in 2010. For every new discovery institutes publish, universities publish about three more.[10] Institutes produce high-quality and often cutting-edge science, yet for every high-impact article from institutes, universities publish two in journals at the same impact level. Institutes expand scientific inquiry and collaborate with scientists across the world, acting as catalysts, while universities publish on a broader array of scientific topics, and their scientists began international collaborations earlier in history and now do so more frequently. And although institute scientists win Nobel Prizes, many more German university scientists have historically been named Nobel Laureates.[11] As shown in Figure 6.1, institutes have steadily increased the number of published STEM+ papers since 1950, but papers from the research universities still make up the large majority of the country's production. In some ways this might not seem unusual; after all, there are approximately three times more university scientists than institute scientists. But this is not what the dual-pillar policy envisions and reinforces—universities are supposedly supported mainly to educate scientists. Yet while universities prepare each new generation of scientists, they also remain the main engine of the country's scientific research, however underappreciated. And given the institute sector's superior financial, recruitment, and working conditions, excellent performance can be expected.

These trends suggest several things about our broader argument. First, the university-science model is not the only one behind science, nor did it necessarily have to be the main one. As this case illustrates, nonuniversity institutes can be successful, and with some modification, they have been replicated outside Germany. Examples include the internal research laboratories within the U.S. National Institutes of Health or the researchers and their laboratories of France's *Centre National de la Recherche Scientifique* (CNRS), or even units of the national Academies of Science in the former German Democratic Republic, the former USSR, or China. While the German case is perhaps extreme in its implementation of a genius model and its resource allocation implications, it represents a long tradition in thinking about the organization of capacity for research, the logic of which continues to appeal. In the U.S., for example, through ear-

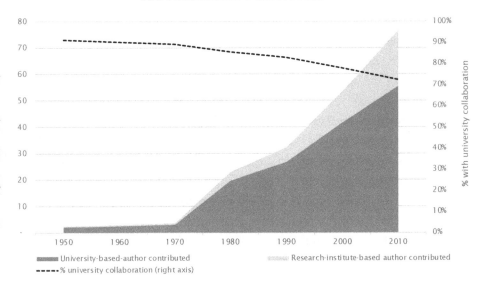

Figure 6.1. STEM+ Articles by Germany-Based Authors in Research Institutes and Universities, and Percentage of Articles from Universities, Germany, 1950–2010. Source: SPHERE database.

lier parts of the century of science, a recurring countertheme lobbied for focused funding mostly to support the "best and brightest" scientists in dedicated environments, often with reference to what was happening in European nations, essentially the Harnack principle; yet the university contribution, while rising via collaboration, continues to vary across top-producing countries.[12]

Second, while no doubt successful, stand-alone institutes are a costly alternative, and if the organization of research had been primarily based on them the overall dimensions of mega-science across the world would have been limited. The German example is telling in this respect. Personnel counts and resource flows are challenging to reconstruct, especially for the diverse research institutes with considerable (and increasing) base funding.[13] With approximately three university scientists for every institute scientist in 2005, we estimate that the annual publications per scientist, including all research personnel, at the Max Planck institutes is about three-fourths of a paper, while for universities the ratio is a half of a paper per scientist per year.[14] This suggests that scientists at most productive institutes are about a fourth more productive than those at universities, understandable given that the former can dedicate all their time to research. Further, using the higher MPI rate of production, for institutes to be the sole producer of the approximately thirty-two thousand publications from universities in 2005, institutes would have required about two-thirds more scientific personnel and infrastructure budget, a very costly proposition. As demonstrated by the research university model's development in the U.S., subsidizing research by expanding postsecondary education is a key component that lowers the full cost of doing science. It is doubtful that even the wealthiest countries could have funded institutes to the degree it would have taken to produce the amazing flow of publications that now exists, chiefly from universities. Also, as will become evident when we visit recent stunning research growth in East Asian countries, an opening up, even a democratizing, of science to a wider segment of society across universities has been a major boost to overall production.

Last, the counterfactual case of institutes indicates how prolific organizations under the university-science model are—even when suboptimally supported. German universities have been productive despite being partially deprived of the full benefits from a deepening education revolution. Despite comparative underfunding and a crisis of legitimation, they have formed the bedrock upon which Germany reestablished so much of its scientific prominence in the highly competitive European and global science systems since its nadir of 1945. By the 1980s—well before the much-touted emergence of Chinese and other leading science producers in East Asia—it was European-based science that reemerged to rival U.S. domination. German universities were fundamental to this revival. Even in a country with arguably the world's best-organized, funded, and prestigious system of independent research institutes, research universities continue to provide the majority of publications and play a major role as a pathway to mega-science trends.

The key reason for this is the robustness and versatility of the form of the research university that Germany was so instrumental in establishing. Counter to earlier predictions of the decline of universities in new science production, this model has proved very durable and is at the heart of journal publications from all world regions. Research universities remain the world's core organizational form for providing the platform for exchange among members of all other organizational forms. They are dynamic and autonomous, benefit from intergenerational exchange and the ambitions of young scholars, and connect the broadest possible array of disciplines, all within an integration of higher education into the broader culture. Institutes may be less flexible and nimble in a changing research climate. In West Germany over the 1960s, as enrollment levels swelled with the education revolution, research-active universities were being founded *en masse*, so that by 1964 11 new higher education institutions had opened their doors. The early 1970s saw the most dramatic expansion, with 46 new universities founded by 1974. Even though East Germany did not generate many STEM+ papers in Western journals, after reunification in 1990 its growth in PhD-granting research universities increased again, while its Soviet-style Academy of Sciences was dismantled.[15] By 2010, re-unified Germany had 118 research universities, a third as many as the U.S. with a fourth of the population. Reestablishing universities based on the principles of their forerunners' integration of teaching and research turned out to be crucial for Germany's significant participation in mega-science.

Over the long century of science, country after country has boosted its science capacity by creating more universities based on the German-Humboldtian ideology that integrates teaching with research, then increases enrollments, hires more faculty, and improves research environments in these universities. This is the secret behind the amazing, sustained world explosion in new discoveries over the twentieth century. Ironically, while Germany provided the world with this innovative Humboldtian ideology and its enactment in the university-science model, in recent decades it has not supported its universities' research capacity at world-class levels—a consequence of the dual-pillars policy and the myths it sustains. Germany will have to do more for its universities soon if they are to remain scientifically productive at the highest level.

SCIENTIFIC RESEARCH EVERYWHERE

Revealing the importance of the university to science production in Germany does not diminish the role of research outside the university's walls. It is a significant source of science, growing over time. As noted before, this confirms part of the earlier Mode-2 science prediction that greater numbers of organizational types would become involved in research over time. In Germany, the number of nonuniversity organizations, including research institutes in particular, publishing at least one annual STEM+ paper more than doubled from 1980 to just over four thousand by 2010. These nonuniversity organizations contributed to about a third of all papers in 1980 with at least one German-based author, and rarely in collaboration with academic university-based colleagues (Figure 6.2). This changed significantly over forty years, as papers solely from nonacademic scientists, such as those in industry, generally dropped, while collaborations between nonuniversity and university-based scientists exploded. By 2010 about 45 percent of all German-authored papers included a scientist working outside of academia, yet with rising collaboration, about 40 percent of these also included university-based scientists. Even though the institutes have been slow to collaborate with universities, due in large measure to their spatial autonomy, specialized topics, resource base, and prestige, other organizations have not: the collaboration patterns are significantly discipline-specific.[16] And numerous funding lines and thematic programs—at national and supranational level—support larger and more diverse collaborations, such as Clusters of Excellence and Collaborative Research Centers, also between types of research-producing organizations,[17] even as "projectification" across scientific fields emphasizes—or even requires—specific forms of research collaboration.[18] Research governance in Germany exhibits considerable state control, albeit decentralized, leading to considerable within-country competition, a major goal of recent policies such as the Excellence Initiative, now called the Excellence Strategy.[19] This policy supported differential types of collaboration and had impact beyond the universities that shared the modest amount of €4.6 million (2006–17), scarcely 4 percent of the total research funds of German universities, as well as unintended consequences of subsequently lowering government funding for those universities that "won" the funding line "developmental concepts" but also fostering collaborations beyond the selected organizations.[20]

Indeed, researchers and the organizations in which they work are subject to increased competition on multiple levels—local, regional, national, and global—yet simultaneously they seem to respond to these multilevel

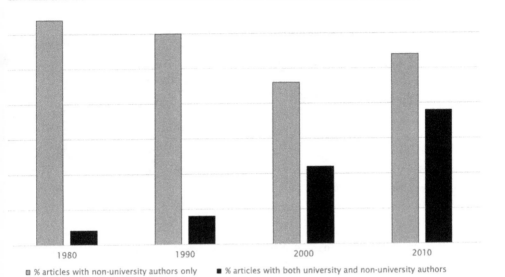

1980   1990   2000   2010

▣ % articles with non-university authors only   ■ % articles with both university and non-university authors

Figure 6.2. STEM+ Articles from Only Nonuniversity Institutions Decreasing and Collaborative Articles from Nonuniversities and Universities Increasing, Germany, 1980–2010. Source: SPHERE database.

challenges with increasing or enhanced collaboration across disciplinary and organizational boundaries.[21] Also in Germany, competition and collaboration are intertwined phenomena that affect research organizations, disciplinary networks, and scientific development and (multi)disciplinary patterns in conducting research, publishing findings, and citing published work.

In fact, due to the increasing ease of collaboration, the number of authors proliferated—on each paper and in general. Coauthorship drives the continued exponential growth in papers published in ever more journals. Furthermore, some authors have multiple affiliations that inflate the number of organizational contributions to papers and large-scale collaborations (see Chapter 9), with more articles published in "hyper-authorship" with over a hundred contributors.[22] In Germany, the number of authors per paper has risen steadily and quickly since 1980, from 1.1 to 2.4 by 2000 and doubling again by 2015 to 4.9, with the latest year, 2020, witnessing the mean authors per article in the STEM+ disciplines at 5.8 (both national and international coauthors).

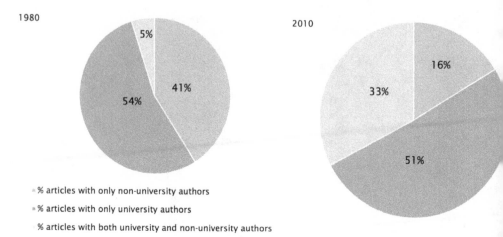

1980

5%

41%

54%

2010

16%

33%

51%

▪ % articles with only non-university authors

▪ % articles with only university authors

% articles with both university and non-university authors

Figure 6.3. Shifting Share of World's STEM+ Article Authorship from Universities Only and Nonuniversities Only to Collaborations across Organizational Forms, Germany, 1980–2010. Source: SPHERE dataset. Note: Size of charts reflects increasing volume of STEM+ articles from 1980 to 2010.

Research collaborations may be increasingly international, but most remain nationally and institutionally embedded, as the conditions for cooperation remain strongly national. Germany's twentieth-century history emphasizes the dramatically altered opportunities for cooperation, both negatively (World War II and the splitting of Germany into two states in 1949–89) and positively (German reunification in 1990). The latter again strengthened intranational research collaborations because the German research landscape was restored, enabling already existing collaborations to flourish and creating new opportunities for collaboration across former borders. However, the transformation of collaboration opportunities results not only from historical events in national politics but also from gradual change dynamics, as most collaborations start from the bottom up, or due to such innovations as digitalization or the diverse research funding instruments with different time horizons designed to stabilize collaboration.

Such trends in competition and collaboration are found in Germany and globally. Instead of a receding role of the university in scientization, as was dramatically predicted to much publicity at the time by Mode-2-science proponents, a two-part change occurred worldwide. Papers authored only by university-based scientists make up about one-half of all papers, a proportion remaining steady from 1980 to 2010 (as displayed in Figure 6.3). The first change is the significant decline in the share of publishing exclusively among scientists working outside of academia, dropping from 41 percent to 16 percent of the world's total in thirty years. The second change is that, as the total number of papers exploded, collaboration with universities did too. By 2010, papers involving collaboration between university scientists and scientists from a growing number of non-university organizations increased from 5 percent to 33 percent of the world's total. These partnerships across boundaries of organizational forms have added significantly to new discoveries, providing a correction of the Mode-2 hypothesis and giving further support to the argument that mega-science relies chiefly on the forum and platform provided by the research university.[23] This is because of the spread of the education revolution and the university-science model well beyond Europe and North America. This general trend is particularly well illustrated by East Asian countries, to which we turn next. Japan and to an even greater extent South Korea, Taiwan, and China have followed this model, in record speed, implementing expansive higher education to educate their citizenry—and developing their capacity for scientific knowledge production.

CHAPTER 7

It's Simple Engineering

Pursuing the World-Class University in East Asia

As global mega-science heated up in the 1990s, a handful of Asian nations entered global science production and made a media and science policy splash in the U.S. and Europe. As described before in Figure 1.2, by 1980 European and American university-based scientists and their nonuniversity colleagues were authoring the lion's share of all STEM+ papers at about 80 percent. In the ensuing thirty years, scientists in Asia and other regions of the world steadily increased their contributions to the world's total output. In many ways, Asian universities' ascent to STEM+ research followed the same pattern found in Europe and then the U.S., but in an accelerated fashion. Until 1950, the number of annual papers from scientists in all Asian nations was negligible, and even in 1960 this region contributed less than 5 percent to the world's total. Then, within twenty years, Asia doubled its share of exploding world growth so that this region accounted for 20 percent of the world's papers by 2000. In 2010, Asian countries produced more papers than the entire world had in 1970, and for the first time surpassed the U.S.'s total output, pulling the global center of gravity of mega-science in their direction. As in the U.S., over 80 percent of papers from newcomer Asia are authored by at least one university-based scientist. And perhaps most important, this region's rapid, condensed experience with building platforms for research around the university-science model shows both the cultural reach of the education revolution and its, at times unintended, impact on mega-science.

The pulling of the world's science center of gravity away from North America and Europe and towards Asia was hardly celebrated in all quarters. Repeatedly asking if American dominance in science production is vanishing, the U.S.'s National Academy of Science and National Science Foundation, and the science policy community more generally, expressed competitive concerns in media, books, and commissioned studies describing the decline of American science and the rise of Asian, particularly Chinese, science.[1] And the fact that recently released STEM+ publications counts show that in 2019 scientists in China authored slightly more than those in the U.S. will only reinforce the concern. Echoing an older specter of an intensely competitive world arising from the economic success of Japan in the 1960s, then from South Korea and Taiwan, two of the so-called Asian Tigers, and most recently from China, the stage was set for the realization that these nations were also making significant contributions to scientific and technological discovery.[2] It is a new take on a familiar story: Asian nations are presumably beating the U.S. and the West at their own game. With some familiar Western bigotry towards Asia added in, it is also implied that these nations are not quite playing the game by established rules. As with earlier reactions to Asian economic success, attention is placed on exotic features of these nation's science systems. Media ran stories on private corporations, such as Samsung, founding a few universities in South Korea; reports of huge cash bonuses paid by the Chinese government to scientists for publication in top science journals; and outlandish salaries to entice famous expatriate scientists to take positions back home. So too, much is made about the fact that some of these presumed "strange" practices might infect other nations. When such tactics backfire, as in the case of fraudulent results by a disgraced Korean scientist who fecklessly claimed he was blinded by the desire to improve his country's scientific stature, they capture intense global attention expressed with detectable distaste for these crass policies.[3]

But such exotica are far from the main story of how a handful of Asian nations and their universities launched themselves into the center of mega-science. What is most central is their incorporation of the same German-European and American-modified university-science model driven by expanding higher education opportunities for wide swaths of the youth populations in each nation. Yet, with angst over which nations are "winning the science game" and with the media coverage and hoopla over highly publicized rankings of universities on various "world-class university" criteria, this continuation of the same process in a new region goes mostly unnoticed. It is not, as many might assume, a story of overtly copying the most prestigious of American universities, although this was tried. Rather, similar influences arise from the same dynamic of forces from the education revolution intertwined with the growing scientiza-tion of society. Looking past the national twists and turns, a remarkably consistent story emerges about how educational development fostered the overall rise of East Asia in mega-science.

First, the rapid development of mass basic education followed quickly by secondary education and then postsecondary opportunities are the foundation from which all else flows. Accompanying these has been a profound cultural transformation of the meaning of the university in these societies, supported by a deepening awareness of research capacities linked with images of national progress. Next, enhanced research capac-ity emerged from intensive participation of Asian scientists in American and European networks of universities (collaboration networks in uni-versities worldwide are described in detail in the following chapter). For example, universities in countries such as Malaysia, Singapore, and Hong Kong have taken a three-pronged approach of nurturing local talent, at-tracting foreign talent, and repatriating diasporic talent from leading doc-toral programs.[4] Finally, this all culminates in government interventions to engineer a few world-class research universities that seemed naïve at first glance but ironically yielded more research-active universities, just as the U.S. did earlier. Although scientists in universities located in twenty Asian nations now contribute papers annually, the history of the process is best told around the four highest-producing nations of China, Japan, Taiwan, and South Korea.[5]

Radical Human Capital Development to an Advanced Schooled Society

Attempting to recover from World War II in the case of Japan, from pre-war Japanese colonial domination in the case of Korea, and from civil war and major social change in the cases of Taiwan and China, these countries would embark on a condensed version of the twentieth-century recipe for development: investment in human capital of populations through formal mass education. Consider South Korea. Oppressed by Japan early in the century, pre World War II Korea had a low literacy rate, limited vocational secondary training, and a small, educated elite, including very few educated women. At the end of the Korean War (1950–53), the forcibly divided, agrarian nation of the Republic of Korea (or South Korea for short), with a devastated infrastructure, a low GDP, and few extractable resources for heavy industry, turned to radical investment in human capital through formal education. Initially employing an efficient, low-cost strategy for provision of schooling, South Korea made primary school universal and opened academic secondary schooling to increasing proportions of students, male and female alike. Building on this, the country then grew access to higher education in a fraction of the time it took the U.S. and European countries to do the same previously. For example, higher education enrollment rates went from about 10 percent of each youth age-cohort to an astonishing 80 percent in just three decades, from 1970 to 2000. Currently, it enrolls 90 percent of youth in some form of postsecondary institution and funds all levels of education on par with the wealthier U.S. and above many other high-income countries. And this rapid development has come with high quality, as South Korea consistently places among the top scoring nations on international assessments of secondary school achievement in mathematics, science, and language skills.[6]

A similar, condensed version of the education revolution occurred in the other Asian science producers as well. Japan grew enrollments in the university sector over the 1970s, from about 20 to 30 percent of all youth, then reached 65 percent by 2015. Taiwan grew upper secondary attendance from just over one-half of youth to nearly all youth in the fifteen years from 1975 to 1990. China's political and social upheavals and its sheer size put it on a slower course, but once started, postsecondary enrollment grew from about 5 percent in the early 1990s to just over 50 percent of all youth by 2015.

Compared to the nearly two-hundred-year journey of the first countries experiencing the education revolution, Japan, South Korea, and Taiwan, later joined by China, rode a fast-paced version, a pace that primed these societies for rapid changes in their universities and then their contributions to science. In the 1980s, South Korea had just 16 universities with scientists publishing STEM+ papers; building on expanding secondary and postsecondary systems, the number grew to 143 universities by 2010, 85 of which were research intensive. Over the same period, Taiwan, with half the population of South Korea, expanded from 23 to 153 universities contributing to research. Japan, which already was producing research in some 172 universities by 1980, added 30 more over the next thirty years. The latecomer massive China had only a handful of universities in 1980, and even by 2003 only 9 of its universities produced enough science to be included among the most research-active universities in the world ranked by the "Shanghai ranking." Fifteen years later, Chinese universities made up over 12 percent of the 1,000 research universities in the Shanghai ranking, and the country had mostly shed the former Soviet research institute model of science production that attempted to relegate universities to more training and less knowledge production.

Not all universities in these nations are equally research-intensive, and many are newcomers to scientific discovery, but the growth in university-based science capacity is impressive. It took the U.S. almost a full century to expand postsecondary education to 70 percent of youth and nearly sixty years to grow a full array of research-active universities. These nations accomplished both in about half the time. Nevertheless, an advanced version of the schooled society and its highly productive research universities takes more than just a rapid expansion of education. It also requires a significant cultural shift in support of the relationship between universities and the greater society.

DECLINING MANDARINS, RISING UNIVERSITY-SCIENCE MODEL

In Seoul's impressive, sprawling, fifteenth-century Changdeokgung Palace sits a lush, formally sequestered garden curving around a tranquil lily-pond. Off to one side in a commanding pavilion, former Korean kings would observe annual examinations in which the best and brightest young men were selected for the kingdom's administrative service. On the other side of the garden is a temple-like building that once housed the king's private library, serving as a kind of research center for the examination-selected political and administrative experts to develop their statecraft. The garden of the palace is now just a tourist site, and visitors from beyond Asia tend to have little idea why an empty lawn holds the rapt attention of the nodding and pointing Koreans, Chinese, and Japanese tourists standing among them. For them, Changdeokgung's garden symbolizes a cherished three-part cultural ideal that was once at the heart of education in many Asian countries. First, administrative elites were to be selected meritocratically through a type of academic examination, albeit from a small set of young men; second, selection through this examination represented a unique royal charter; and last, the selected were expected to lead and solve problems of the collective kingdom. Intelligent, Confucian-steeped gentlemen holding a similar model of the perfect society shared through a universal, high-form communicative style with a deep sense of officialdom was the goal of the examination and its selected applicants.

The first Asian imperial examination blossomed somewhere during the seventh century in the Chinese Tang Dynasty, more than a half-century before the first fledgling universities were established in Europe. A form of nonhereditary elite selection, the examination system spread throughout eastern Asia. Dubbed "the Mandarins" somewhat inaccurately by Western intellectuals, these men were revered in their own cultures, and represented an early merging of advanced education with high social status. It would take many centuries before the examination, selection of scholar-officials, and the king's library would be rolled into something similar to a Western university. Becoming dysfunctional and anachronistic, the formal Mandarin process would officially end in the early twentieth century, but some of its ideology, nevertheless, remained robust and pervasive as a cultural form. Regardless of imported cultural ideas and hegemonic influence from the West, the Mandarin ideology would continue to define the relationship between Asian universities and their societies well into the twentieth century. Universities in South Korea, Japan, and Taiwan became the new selectors and verifiers of a kind of modern Mandarin. With the dominance of the Chinese Communist Party, a radical rejection of the ancient Mandarin system, and a focus on Soviet-style state research institutes, universities in China did not follow the same course over the twentieth century, although recently they have engaged in major research production through partially adapting a university-science model.[7]

Of course, there are prestigious universities in all nations, but in some Asian countries a small number of their universities are revered beyond what would be the case in the West. Successful performance on entrance examinations to a top university in Japan and South Korea, for example, comes with direct connection to the best starting positions in government and the private sector, without much need to further prove oneself during the pursuit of a first degree. And unlike in the U.S., without entrance to one of a clearly identified set of universities it is nearly impossible to enter positions leading to elite status in the society. Hence, the equally celebrated and condemned "education fever" manifested in all kinds of expensive preparation for examinations to gain entrance to these elite-forming universities is a direct outcome of the early relationship between advanced education and the Mandarin legacy. In South Korea, "SKY," the widely known acronym for the three high-status-confirming universities of Seoul National, Korean, and Yonsei, is synonymous with selection into the nation's elite. The ideal of assumed meritocratic selection by examination, verified by a circumscribed set of universities and chartering to adult elite status, still permeates the culture.

While a modified Mandarin culture continues, the forces behind greater human capital development and an advanced schooled society steadily caused a rival model of the university within society. Well beyond elite selection and special skills for the relative few, the newly accessible university broadened into many parts of society just like it had in the West. The elite status of some universities remains, but the significant growth in newer, and in most cases less-prestigious, universities steadily moved towards a similar university-science model already operating elsewhere. Nevertheless, a misunderstanding of what really drove the American version of the research university, plus a lingering Mandarin ideal, led these governments to try to spend their way to a small set of highly prestigious and productive research universities. Although these programs failed in one sense, the effort ended up leading to far greater research capacity and widespread production than ever imagined.

BUYING A WORLD-CLASS UNIVERSITY

By 2013 about two dozen governments across the world had attempted to take extensive sums of funds and aim them at a selected few of their best universities to stimulate more science production. In every case the unabashed goal of these "excellence initiatives" was to create a "world-class university" for their country, meaning an exceptionally productive research university that could compete with the most scientifically pro-ductive in the world. In the mid-1990s, when the first of the excellence initiatives were launched, the Chinese and British media rankings of the world's best universities made two points abundantly clear. First, a set of very prestigious institutions were producing more scientific discovery than the average university, although this fact was exaggerated in the thinking behind the initiatives. Second, since many at the top of the world rank-ings were well-known American universities supported with vast amounts of openly celebrated resources, the conclusion was that money must be the way to excellence. For example, by the 1990s, Harvard, Johns Hop-kins, and MIT had operating budgets already in the billions of dollars, massively more than most universities elsewhere. Although the national excellence initiatives used somewhat different tactics, a common strategy was to identify a country's best universities and award them large sums of extra funding to do more and better science. Unselected universities and their potential contributions to science were not part of the strategy. The idea was to spend money to engineer what is called in sociology of science *cumulative advantage*—make the richest centers of science even richer, and hopefully more globally prestigious.[8]

At the time, this seemed the obvious way forward. The few very wealthy American universities at the top of world rankings were happy to take the credit for intensifying mega-science, even though, as already shown, it was more than money in the accounts of a very small set of universi-ties driving European and American science production. Overlooked in the rush to embrace the excellence initiatives strategy was the underlying fundamental platform provided by the university-science model—widen-ing of undergraduate education, cross-subsidizing research capacity, and an opening up, even democratizing of sorts—or at least deemphasizing of prestige—in the science capacity of many universities, often based on international recruitments and collaboration.

The public and governments of the four soon-to-be-high-producing East Asian countries were ripe for such excellence initiatives, and each pursued one enthusiastically. Their postsecondary systems had rapidly ex-panded; the older mission of Mandarin selection was partially supplanted by a mission of new knowledge creation and participation in the globalization of science.[9] Ministries of science, sometimes along with ministries of education, developed bold, futuristic-sounding programs: Taiwan's World Class University Project, China's 985 Project, Japan's Center of Excellence in the 21st Century (COE21), and South Korea's Centers of Excellence (CER) and Brain Korea 21 (BK21). It was thought that what was needed was to let these programs work their funding magic and engineer a Stanford, a Johns Hopkins, or a University of Michigan to be the nation's world-class university (or a small set of such) at the pinnacle of science production. Fortunately for the future of science in these nations, something unexpected resulted instead.

There are now systematic evaluations of the impact of some excellence initiatives that show what happened. Taiwan's World Class University Project, or WCUP for short, is a telling example.[10] Out of the country's 145 universities at the time, about 30 of the most research-active universities were invited to apply. Then 10 of the best of the best were selected and awarded extra funds totaling US$1.5 billion from 2005 to 2010 on top of their regular R&D share, an augmentation in the most generous cases of 25 percent more research funds. As hoped for, compared to the period before the initiative, the funded universities did indeed increase their output rate, including publication in journals with higher impact factors. Also as intended, the selected universities' administrators used the funds to increase an array of policies and actions to facilitate their university's capacity for quality science. Quite unexpectedly though, not only did the 10 selected and funded universities increase their output rate, so did the 20 unfunded universities. By the end of the program the latter's rate of increase in volume and impact of papers was greater than the universities the WCUP had destined for world-class status!

This was not the intention. As the Taiwanese case demonstrates, it seems that "engineering" a cumulative advantage is harder than assumed, maybe even impossible in an environment of expanding postsecondary education and the resulting growth in universities and collaborative research. Taiwan's WCUP did not catapult a few of its universities into the top of world rankings. Indeed, the widely acknowledged nation's best, the National Taiwan University, has steadily remained at approximately 150 out of the world's top 200 since the early 2000s, albeit keeping its position in a world of rising capacity. Instead, the WCUP super-charged the environment of all the country's universities with the goal of producing more science. Clearly the unselected universities got the message and felt the growing competition and the weight of new ideas about how faculty and students were to be engaged intensely in scientific discovery. And like their funded competitors, administrators at the unfunded universities also implemented some of the same research-facilitating policies. The WCUP created internal competition and awareness of science production goals, in addition to the motivating idea that a university-science model could be for all postsecondary institutions. This ultimately strengthened the scientific capacity of the entire system of universities.

A similar pattern is reported for other countries' excellence initiatives. From 1998 through 2001, China's 985 Project, for example, invested about US$500 million in the country's top-tier institutions, such as Beijing and Tsinghua Universities, to stimulate more production. And then the program invested significant, but considerably less funds per university, into a group of second- and third-tier universities. While the lavishly funded Beijing and Tsinghua Universities did take a major leap into the top 100 world-ranked universities (to position 57 and 45 respectively), it was the least prestigious universities, which received less funds, that increased their publication rates the most, net of other factors, just as in Taiwan.

At the end of the 1980s, Japan was producing more science publications than Germany and was second only to the U.S. With the globalization of science intensifying, though, its total share had been eclipsed by China, Germany, and the U.K. by 2000. Created to address what was perceived as some stagnation, Japan's COE21 and related programs beginning in 2002 were essentially government-supported excellence initiatives, distributing more funds, again to the selected "best," in an openly stated strategy to create world-class universities. And much the same outcome occurred. While these funds did contribute to further production among the top universities, such as University of Tokyo and Kyoto University, their position in the Shanghai ranking did not change much. In addition, a group of universities with less national prestige, which before the COE21 programs had been the second-largest producers of science publications in the country, went substantially less funded, even though they continued to produce. Tohoku University's international higher education scholar and our colleague Kazunori Shima has concluded that in the early mega-science period these universities were the "unsung heroes" of very high production, which in subsequent excellence programs were unwisely ignored.[11]

As noted, South Korea also undertook a series of excellence initiatives, committing several billion dollars to universities to make them "world class." The funding formula reveals the ideal enactment of accumulated advantage; the size of awarded BK21 funds to universities and their pre-funding level STEM+ publications was nearly a perfect positive correlation, with the highest awards substantially larger than the rest. This, of course, makes it difficult to assess the long-term influences of the additional funding. Unlike in Taiwan, the Korean policy implementation did not create a nonfunded, comparable group of universities acting as a kind of control group for comparison. Nevertheless, the best attempts to examine the impact of Korean excellence initiatives conclude that most likely the additional funding did enhance the science capacity and the production rate of faculty at the most highly funded universities. But the program's goal of placing ten of its universities within the world's top one hundred fell short with only four there in 2018, and all of these were already in this group from the start. And just as in the other countries, many lesser-funded and even nonfunded universities also increased their publication rates.

In each national case, while additional funding unsurprisingly enhanced production at already scientifically strong universities, the unselected or less-funded universities were also motivated to implement a robust form of the university-science model that increased their performance. The excellence initiatives were too mechanical to achieve the much-celebrated goal of lifting a university into the top rankings or quickly further increase the standings of those already there. Yet clearly the additional support played a role in these countries' ability to become major contributors to mega-science. While in a narrow sense these policies failed, they did stimulate a major success story on an extensive scale. The mere presence of an excellence initiative sent a strong message to all institutions that a new model of the university was ascending. For example, in Korea, so many teams of research administrators from scores of less prestigious universities attended the ministry's announcement of the new competition for science centers that it created a temporary hotel room shortage in Seoul. With each subsequent wave of CER and BK21, more universities were funded, although some by modest amounts, but the subtle message from the beginning was clear: "Your university too can join in science production." Or perhaps more to the point, your university should and must join in to be viable into the future. In Taiwan's case, centralized, low-tuition policy means that public and private universities heavily rely on governmental appropriation and subsidies for operation and hence pay close attention to national initiatives. And, selected or not, all the country's universities got the message that doing more research is central to future viability. A kind of aggregated "publish more or perish" edict pervaded the environment of the excellence initiatives, and this led to new organizational behavior. For survival, organizations must be adept at conforming to new norms of operation, and this extends to the university-science model. As the evaluation of Taiwan's WCUP shows, better research facilitation practices spread from funded to nonfunded universities, and a version of this likely occurred in all these countries.

Not only were the excellence initiatives too mechanical to succeed in the narrowest sense, ultimately, they were a misunderstanding of what drove the university-science model in the first place. Funding resources are important, of course, and certainly the science production of selected universities is better because of increased funding, but the impact on changing the culture of all universities was underestimated, if considered at all. Ministries intended to fund their way to an MIT or any of the U.S. Elite-16, but instead they got a full set of widely scientized universities collaborating with each other and many other research universities around the world. And this, of course, is very similar to what occurred in the U.S. and increasingly elsewhere. World-class universities are fine to wish for, but a few are hardly enough to sustain any one country's participation in the intensive global mega-science production, based on global networks among scientists working in universities worldwide.

Also important, a policy ostensibly narrowly aimed at science was, in fact, deeply supported by a rapidly unfolding education revolution. What can be called an "advanced schooled society" is one in which, like South Korea, Taiwan, and Japan (China is not there yet, but is progressing in this direction), almost all youth can expect to attend some postsecondary education. As education expanded in the West in the 1960s and 1970s, a full worldwide expansion was considered unlikely. Perhaps many of the world's youth would eventually be able to complete secondary education but going to college and universities was still assumed to be something for only a minority of populations. Since then, the culture of the education revolution has steadily swept aside an earlier vision of a differentiated society by education level, in other words, stable segments of mostly secondary educated and a smaller university-educated elite. Instead, education tends to stratify society in a more dynamic way. As Talcott Parsons anticipated, an advanced schooled society in which the university plays a major part in the general culture is a new societal form. This is the most instructive lesson to take from what has occurred to universities and science production in Asia. All the handwringing over exotic research practices pales in comparison.

South Korea: University-Science Model in an Advanced Schooled Society

In addition to an excellence initiative, Korea's overall experience with the education revolution and science offers a coherent observation of the consequences of a spreading university-science model. Also, because in some ways the country has recently surpassed the U.S. and Europe in the expansion of advanced education, it is a particularly informative case from which to consider its sustainability into the future. In addition to being at the leading edge of a schooled society, it also leads the world trend towards a declining population, which is spreading throughout the Northern Hemisphere with major implications from looming constriction of advanced education and reduced numbers of universities and their associated research capacity.

As already described, South Korea epitomized the robust pursuit of mass education educational development, going from a mostly illiterate population in 1950 to being among the top ten contributors to the world's science publications fifty years later. By the early 1980s, an expanded secondary system was beginning to create significant demand for postsecondary education, and this expansion would provide the country's foundation for science. In the 1980s, at the beginning of its rise in research, only faculty at the three prestigious SKY universities were publishing more than the occasional STEM+ paper. By the late 1990s, faculty at twenty more universities were publishing at a rate equal to or greater than the SKY faculty had a decade before. Yet just two of the SKY universities still accounted for more than half of all papers with Korean university-based scientists. But by 2000 this had substantially changed, as the SKY universities published less than 30 percent of all papers, and notably over thirty other universities were making regular substantial contributions to the countries' science productivity, as conservatively measured in the leading journals. This trend continued, so at present no one university publishes more than 10 percent of the total and seventy-nine universities annually publish over a hundred papers, while twenty-four regularly publish at least a thousand papers a year. So too, the network of collaborations among scientists across universities (heavily BK21-funded or otherwise) rose over the same period from virtually nonexistent to a thriving, dense network by 2016, and the proportion of papers in high-impact journals is similar across the country's universities.[12]

The Korean case has been significantly shaped by consequences of a lingering Mandarin ideal combined with the national policy of unabashed human capital formation: an elite form of education meets mass demand for postsecondary education. The Mandarin ideal rests on the widely held notion that high-stakes academic examinations are a fair, even meritocratic, selection process, and Korea employs them and a set of very advanced educational policies to make public education up through secondary school perhaps the most equitable—and with the highest-quality academics in the world today. This is evidenced by the fact that Korea has one of the lowest indicators of family background's influence on academic outcomes while perennially being among the best-performing countries on international achievement studies. The flipside, and a consequence of this enviable education development, is one of the highest uses of private supplementary education services, such as tutoring and cram-schools, known in short as "shadow education." This creates a dialectic of opposing forces that pushes each—equitable, high-quality schooling *and* growing private investment in educational preparation—forward. This dynamic also has pushed a growing

demand for postsecondary education. The very public rankings and associated occupational and prestige payoffs across a public hierarchy of universities anchors this process, and as higher education grew from a smaller
elite affair to mass proportions, the shadow education practices once used
only by children of the elite spread throughout society. As a once limited,
yet still revered, process became increasingly open to more parts of society,
an opening up of the Mandarin logic for everyone swept large proportions
of families and youth into considering universities as necessary for their future. Even though it was unrealistic to think that most youth would make
it to the SKY universities, the spreading of this potent cultural form reinforced the motivation for wider participation in postsecondary education in
general. A strong cultural force along with the ever-growing technical and
advanced service portions of the economy sent enrollments climbing, and
along with them participation in university examination preparation. As
secondary and postsecondary education grew over the 1960s and 1970s, a
mere 15 percent of mostly elite families were annually purchasing almost
US$1.1 billion (constant dollars) in shadow education services—aimed at
gaining admittance to a SKY university. As universities grew in number, so
did shadow education, which is now purchased on average at an amount
representing a tenth of a large majority of families' budgets.

Somewhat ironically, as more families and their children competed for
places in higher education, the society at large supported the notion of
supplying just the best universities with greater funds to achieve a world-
class distinction through its excellence initiatives. Regardless of whether
their children stood much of a chance of acceptance into a SKY university,
public support came from a wish to reify on the world's stage what was
already so assumed within Korean culture—its prestige ordering among
universities. Similar support from a newly college-educated populace also
acted as an early legitimation of Taiwan's WCUP. The Korean version of
the education revolution was driving greater postsecondary participation
and in doing so generated a wider public interest in affirming—or essentially reaffirming—the old Mandarin ideal, and perhaps ensuring the prestige ordering among South Korea's expanding number of universities even
considering inevitable mass enrollments. As much as funding mostly the
best could be considered an unfair treatment of unselected universities, in

Asia, Europe, and elsewhere, excellence initiatives were mostly an accepted strategy by populations undergoing advanced educational development.

South Korea also wrestles with one of the lowest birth rates in the world. The country recently reported a rate of less than one child per woman of childbearing age, well below a stable population replacement level of two births per woman. Already some elementary schools have more teachers than students, and schools and a few universities are closing. The government and society are actively debating the long-term implications and what can be done about them, including within the higher education sector. So too, as South Korea successfully achieved widespread postsecondary education, a shrinking population meant there will soon be too many postsecondary institutions for ever-smaller cohorts of future secondary school graduates. Korea's BK21 program tried to address this issue too. Through building capacity for training STEM+ PhDs and postdoctoral researchers, the initial idea was to lower undergraduate enrollments at the best BK21-awarded universities, freeing them to become "world-class" *graduate* institutions. It was assumed that this would achieve two objectives. First, by taking the SKY and a few other top universities out of the undergraduate mix, it would result in a reduction in the demand for shadow education, always a goal of the government. Second, this would be the start of a downsizing of less prestigious universities to prepare for smaller cohorts of undergraduates. Unsurprisingly, resistance from faculty and an ensuing series of the country's ubiquitous university student public protests lead BK21 administrators to back off the original goals of creating a bifurcated set of graduate universities and undergraduate teaching institutions. What happened instead was the stable funding of graduate programs, faculty evaluations based on research production, and improved research administration across many universities—essentially a reinvigoration of the strategies of the university-science model. All of this played a major role in the country's rise in mega-science. Exactly where declining cohorts of youth will take this country's universities is less clear. Implications of population trends for the sustainability of the university-science model and research production in places like Korea and the world in general will be considered at the conclusion of the journey.

The speed at which the education revolution unfolded in a number of Asian countries, their adoption of variations on the university-science model, and their ensuing dramatic rise in the production of science at their universities attests to the globalizing process behind STEM+ research. This was all done in the face of notable differences among countries' historical paths and the region's lingering ideology about the role of the university in society that was decidedly different from the West. It is a remarkable convergence and dissemination of ideas and their practical implementation, yielding what theories of large organizations refer to as "isomorphism," an isomorphism now spreading to ever more countries, large and small.

CHAPTER 8

Mega-Science Goes Global

Investing in the Twenty-First-Century University

Education City is a mind-boggling place. Perched on the Persian Gulf on nine square miles of originally barren sand dunes in Doha, Qatar's only real city, Education City collects architecturally stunning buildings surrounded by carefully manicured landscapes. These places are maintained by armies of low-income, Pakistani and Indian imported workers toiling in desert sun as part of the country's notorious labor system.[1] Easily mistaken for opulent corporate headquarters, instead each signature building houses the branch campus of an invited, well-known, usually North America–based university. Education City is literally an oasis of knowledge. Qatar bought and imported the university-science model, ostensibly to educate the nation for the coming knowledge society but with an interesting wrinkle: there are hardly enough university-ready Qatari youth or faculty to fill the campus to capacity. Instead, as is done for nearly all jobs in this unimaginably energy-wealthy but impracticably tiny country, highly mobile students and faculty are imported from all over the world. Essentially well-trained and well-paid migrant workers, the faculty are not, however, imported just to teach foreign undergraduates and the occasional Qatari; they are there in this most unlikely of settings to undertake cutting-edge scientific discovery. And they do so, from petroleum engineering to biochemistry to foreign service. Sparing no expense, the Qatar Foundation, a vast development fund controlled by the emir and his ruling family, fully appointed the Education City campus with state-of-the art labs, research facilities, and necessary cutting-edge instruments.[2]

The scope of Qatar's research ambitions goes well beyond its narrow economic national interests in liquid natural gas and date palms. As strange as Education City appears at first, its shining signature buildings for multiple universities are a manifestation of the attraction, and taken-for-granted-ness, of mixing advanced education with science. Certainly, this is how Qatar's government sees it, proudly publicizing its aim to be a "commercially successful research and development (R&D) hub that attracts the best and brightest."[3]

One of these best and brightest is Dr. Najeeb Halabi, a Lebanese biochemist researching the cellular mechanisms of cancer at Weill-Cornell University Medical College at Education City. He personifies Qatar's ambition to be a world center of knowledge production. When asked about working as a scientist in this out-of-the-way place, Doctor Halabi immediately replies that even without prior connection to Qatar, coming to Education City made a lot of sense professionally. Over his faculty career in Qatar, he has had significant access to the same resources needed to undertake cutting-edge biological research found at top research universities in North America or Europe. Initially educated in Beirut, following with a PhD in biophysics from an American university, Halabi also exemplifies the intensive globalized nature of scientific research, including international research collaboration and coauthorships.[4] He is one of a growing cadre of young hypermobile scientists unconfined by national or linguistic borders; the logical next step in the internationalizing of STEM+ training so important to the continued science capacity in increasing numbers of countries. When the American scientific labor market tightened, he looked across the world for opportunities to further his scientific career—and to pursue his research questions. Researcher Halabi does not see his decision to pursue his projects at Education City as unusual or limiting. Instead, he considers the opportunity as a fortunate step in a successful scientific career.

Five flight hours away, the capital city of Luxembourg could not appear more different from Qatar's Education City: quaint, verdant, and cosmopolitan. The Grand Duchy of Luxembourg, nestled between Belgium, France, and Germany, is home to 626,000 people, of which around half are native-born.[5] The country is breathtakingly polyglot, with Luxembourgish, German, and French as official languages and Portuguese and English also spoken by large minorities. Daily, many tens of thousands commute from surrounding countries for education and work, bringing their native language and cultural sensibilities with them. More than half of the workforce consists of cross-border workers from the "Greater Region."[6] With its strong financial industry, demographic growth, and the EU institutions, Luxembourg has among the highest per capita GDPs worldwide, encouraging such tremendous migration flows, as well as rising poverty rates due to sky-high housing and rental prices, offset somewhat by free public transit and government-sponsored day care.[7]

For centuries, Luxembourg has been at the crossroads of European history, with all that implies. It fit seamlessly into its regional context in the heart of Western Europe, except for one notable exception. Unlike its neighbors, who created and nurtured the evolution of the eight-hundred-year-old Western form of the university, Luxembourg did not officially establish its own national research university until 2003, with much earlier attempts by French Jesuits and German academics failing due to resistance within the country's elite.[8] Their children had, for many decades, been sent off to universities abroad, mainly in Belgium, France, Germany, Austria, and Switzerland, with the wish for exposure to Europe's leading universities and for the youth to experience life beyond the Grand Duchy and establish Europe-wide networks.[9] With the rapid growth—in size and reputation—of the University of Luxembourg, or "uni.lu" as it is called for short, this has begun to change. Responding instinctively to European influence and initiatives, entrepreneurial policymakers finally decided that Luxembourg, too, must have a research university within its borders. Since the education revolution had already unfolded here, this decision was less about educating its youth than to extend its economic fortunes beyond steelmaking and finance to diverse fields, increasingly based on scientific research. The promise of technological innovation and economic growth inherent in the research university was, for a dedicated group of policymakers, simply irresistible. But like Qatar, and scores of other nations worldwide, this investment in higher education and extensive research capacity represented largely uncharted territory for Luxembourg. There were influential critics to be convinced. A mixed-use campus melding residential, commercial, entertainment, and cultural spaces has been developed at a cost of over €1 billion. Smartly appointed within the remnants of a mighty steel factory at the core of the European Union's precursor, the European Coal and Steel Community, the research university also poses a risk for Luxembourg. Can such countries with their new universities compete with those whose institutions of higher learning have had centuries to successfully build their infrastructures, nurture generations of graduates, and break new scientific ground that established their global reputations?

Qatar and Education City's over-the-top reach into higher education with a native population no larger than a regional city and just a half-century removed from a nomadic lifestyle, along with its aspiration to be a global hub of scientific research, appears a huge gamble. A similar sense of a risky wager hangs over the creation of uni.lu in a small country that had met its higher education needs for centuries without a full-fledged flagship university of its own. The same can be said about Najeeb Halabi's bet on a scientific career in a place that before a recent World Cup most could not locate without Google maps. Yet whether Qatar can achieve its lofty knowledge society goals, or if highly developed Luxembourg should have entered the science production game so late, or if talented scientists should venture to these new universities are not the right questions to ask to understand what is happening.

The key to the puzzle that is Education City, uni.lu, and an increasingly globally mobile scientific labor force is to understand that the same university-science model powered by the education revolution's ideology has over recent decades shifted into an accessible worldwide research-producing machine. And universities and their scientists can be fully involved from almost anywhere. The gambles, institutional and personal, may not be as great as they initially seem. Now, hundreds of thousands of scientific journal articles per year are the product of scientists working collaboratively, more than ever, from diverse locations worldwide—including Luxembourg and Qatar, once poorer countries respectively known mainly for steelmaking and pearl harvesting. Today, both countries invest billions in advanced education and science, plus an array of supporting organizations mostly with the same logic perfected over the century of science. Influenced by the inspired leadership of a powerful woman in each case, the details behind these countries' stories are as different as their climates, yet they both demonstrate how the university-science model has become fully taken for granted—and adaptable across the globe.

Two Formidable University Founders

Moza bint Nasser was born in 1959 in Qatar. Her father, a well-known opposition activist who led his Al Muhannada clan into self-imposed exile to Kuwait in 1964, returned in 1977 when Moza married Hamad bin Khalifa Al-Thani, who became emir of Qatar. She later studied sociology at Qatar University and public policy in Islam at Hamad bin Khalifa University's Faculty of Islamic Studies at Education City. Attractively blending traditional and Western elegance with an anachronistic commanding presence of a veritable queen, Sheikha Moza has for several decades been instrumental in guiding a host of education and social reforms that have utterly and swiftly transformed Qatar's capacity for higher education and science. Especially as chairperson of Qatar Foundation for Education, Science and Community Development, she oversaw the flagship project of Education City, the twelve-square-kilometer campus to house Qatar National Library and the branches of renowned international universities invited to bring the world of science to Qatar.[10] Engaged in diverse research, economic, and social development projects, Sheikha Moza served as vice chair of the Supreme Education Council and Supreme Health Council, enacting major top-down reforms of her country's public schooling and health care systems. Among numerous others, the most significant institution buildings whose construction she oversaw are Education City and Sidra Medical Center. Together with the national Qatar University, the international branch campuses at Education City secure talented educational and scientific professionals to work in Qatar's rapidly developing economy. They bring to Qatar diverse curricular offerings related to national needs, especially education, petroleum engineering, journalism and communications, computer science, medicine, and diplomacy. Awarded numerous honorary doctorates for public service in education, science, and human rights, Sheikha Moza was appointed as UNESCO's Special Envoy for Basic and Higher Education and a member of the UN Millennium Development Goals Advocacy Group.

Growing up in the shadow of a steel mill, Erna Hennicot-Schoepges was born in 1941 in Luxembourg to a father originally from Germany and a grandfather whose father had moved to Luxembourg from France, for Luxembourg a not atypical multicultural heritage. Similar familial backgrounds, the small size of the country, and the historically porous national borders confer on the citizens of the world's last Grand Duchy pan-European, cosmopolitan attitudes and global connections. And Hennicot-Schoepges realized ever since her childhood "that there was more than just one country within a Europe that was different from the Europe we understand today."[11] Influenced by the aftermath of the Second World War and musical studies throughout Europe, she went on to be elected to the country's parliament, becoming this legislative body's first female president of a country trading its regional industrial past for a future as a global center of banking and of supranational governance in Europe. Hennicot-Schoepges, with an intelligent, no-nonsense gaze, often stylishly attired in the requisite power suits of Western European officialdom, went on to serve as a cabinet-level minister with a broad portfolio. She became a distinguished cultural entrepreneur who greatly facilitated the transformation of the Grand Duchy's cultural and educational infrastructure, including signature buildings for the Philharmonie Luxembourg and the Musée d'Art Moderne Grand-Duc Jean.

Given her past, it is not surprising that Hennicot-Schoepges, like other of the country's leaders, would start to change her perceptions of the age-old arrangement of essentially outsourcing higher education for Luxembourg's youth to neighboring countries. This started precisely as Europe was making this even more possible, and also at a time when the EU provided member states with greater motivation and rationale to consider their own universities as centers of knowledge production in an increasingly integrated Europe. Hennicot-Schoepges was convinced that Luxembourg should embrace these very ideas and thus continue being Europe's "model student." Fittingly, she was present at the signing of two cooperative agreements among universities in the EU to enhance their competitive standing in the world: the Sorbonne declaration in 1998 and the Bologna declaration in 1999, the latter of which she signed—prior to the founding of the University.[12] Before Luxembourg had even decided to establish a national university, but with brewing interest in such, she took over a rearranged Ministry of Culture, Higher Education and Research, with responsibility for linkages among culture, education, and science.

The process intensified with the launch of Europe's Lisbon Strategy just months later with one central goal—to make Europe "the most competitive and dynamic knowledge-based economy in the world, capable of sustainable economic growth with more and better jobs and greater social cohesion."[13] This demanded that European member states invest more into public research, providing another substantial external impetus to reshape Luxembourg's higher education and research landscape.[14] And within this political and education development environment for Europe in the new century, Hennicot-Schoepges was instrumental in establishing uni.lu, widely recognized as the pivotal leader in the university's creation in 2003.[15] Today, Luxembourg has developed into one of the world's key financial centers, has the highest GDP per capita in Europe, and already by 2010 had the highest proportion of professionals in its workforce among thirty-five developed economies.[16] Luxembourg's hyperdiverse and growing society facilitates its knowledge-based economy with an increasingly highly educated workforce.

FUTURE INNOVATIONS
OF THE UNIVERSITY-SCIENCE MODEL

These two unique university founders indeed transformed their small countries, contributed to their scientization, and built intercultural linkages in their regions reflecting global norms relating to research universities. The universities of Qatar and at Education City and uni.lu are based in different world regions, flourish within contrasting institutional environments, and utilize very different resource bases. Yet, along with older education and knowledge hubs, such as Malaysia, Singapore, and Hong Kong, they are similar in their ambition to rapidly become part of global mega-science and in most ways essentially employ the university-science model.[17] Luxembourg and Qatar are also a new take on the university-science model, one that offers some hints as to possible innovations that could emerge elsewhere in the near future. Each country selected internationalization strategies that rely heavily on cross-border migration and international mobility; a comfortable strategy given their earlier reliance on outsourcing advanced education to other countries. Built with resources from wealth from the private sector and public dedication to university institutionalization, and enabled by extraordinary migration flows of faculty, staff, and students, each has opened its doors widely in recent decades to scholar-scientists and students from around the world. They implement far more explicit strategies than the earlier unplanned internationalization of the STEM+ research capacity and the "stubbornly parochial" social sciences in the U.S., for example.[18] Ostensibly "national" flagship universities, these universities are conspicuously and thoroughly international, remarkably collaborative, and expected to be "globally engaged and competitive"—lending the world's university rankings a new international dimension not captured in the image of one country's universities versus those of other nations.[19]

Of course, universities are all tethered to their national context to some degree, especially through funding, yet they increasingly exist in a networked world. And this expands to a kind of global cross-subsidizing of research capacity through attracting youth from other countries motivated by the education revolution to attend their undergraduate or master's programs. At Education City, there are more undergraduate students from outside Qatar than from within, and uni.lu attracts students from all over Europe and beyond, without which higher education and science in these countries could not flourish.

Likewise, these newer international national universities facilitate the accelerating global reach of higher education and science as they systematically foster the global recruitment, cross-border mobility, and multicultural networks of scientist-faculty.[20] Uni.lu, for example, participates in the University of the Greater Region, a cooperative venture, originally funded by the European Commission, between universities in Belgium, France, Germany, and Luxembourg.[21] Education City hosts prominent foreign universities' branch campuses; Qatar University's own campus is also ultramodern. The University of Luxembourg, once housed in several regal structures in the city of Luxembourg, has moved the bulk of its operations to the *Cité des Sciences* (City of Sciences), the state-of-the-art campus for scientific research built within a refurbished steel factory site in Esch-Belval at the border to France. The Cité des Sciences comprises a dozen new buildings, including research institutes, startup incubators, R&D companies, the environmental ministry, the national innovation agency, and one of the world's fifty atomic clocks. More than ever, these universities are embedded in transnational economic, political, and cultural networks, relying on continuous migration and widespread mobility.

Education City and uni.lu, with their ambitious agendas, operate in very small but exceptionally wealthy countries in which each university system emulates global goals simultaneously while serving national and local needs. Both countries, lacking the decades of development of advanced education that have conferred a massive head start to many in Europe, North America, and Asia, used a blend of extreme wealth and an explicitly internationalized university model to jump start their entry into mega-science. Qatar and Luxembourg are each novel, but even in their unlikeliness they are testament to the worldwide diffusion of the university-science model. Before the establishment of their universities according to the university-science model, scientists in each country were publishing only several dozen STEM+ papers annually, but by 2010 about three hundred were regularly from Qatar and around four hundred from uni.lu and supporting organizations in Luxembourg. While their aggregate output is a mere drop in the bucket compared to the total volume of mega-science, these universities are contributing just like thousands of others worldwide to the common effort of scientific advance—more than ever collaboratively.

Going Global

As shown in Figure 1.2 in the first chapter, several countries outside of North America, Europe, and Asia had scientists involved in about one-fifth of all STEM+ papers in 2010, doubling their annual volume over the prior two decades. Most of these countries are not suddenly super-rich, nor have they necessarily used some hyperinternational strategy, but they all are contributing through universities supported in part from an advance of the education revolution. As noted before, over the century of science enrollment in higher education went from a miniscule fraction of the world's youth population to 20 percent by 2000.[22] Less than two decades later, this proportion has nearly doubled, with more than 220 million youth, or about 38 percent of the college-aged cohort, now enrolled in postsecondary education of some type.[23] In addition to growth from a combination of countries already with established higher education systems, this includes newcomers as well. All of which means more universities undertaking more STEM+ research.

As of the mid-1980s about two dozen countries outside the three major regions of paper production were regularly contributing at least a minimal amount. By 2010 this had grown to universities and other research-producing organizations in an additional 120 other countries. To get a sense of this trend, Table 8.1 displays 22 countries from Oceania, sub-Saharan Africa, South America, the Middle East, and Southeast Asia whose scientists authored at least nine hundred papers in 2010. And while some output levels are modest, some, such as those from Australia, Brazil, Turkey, Iran, and Israel, are more substantial. And in each case, scientific research has grown significantly; the average country here increased its annual total volume of papers about twelve times over twenty years, with some cases of extreme growth. And as expected, except for Kenyan and Tunisian scientists, the vast majority of authoring scientists are university-based.

Also as expected, these countries are engaged in expanding postsecondary education to a broader proportion of their populations, and there is a modest association among them between more enrollment and greater number of STEM+ papers. Observing the history of postsecondary enrollments in other countries indicates that enrollment rates grow fast once about a third of youth are included and then slow once about 80 percent are attending. Many of these countries are entering a rapid-growth period; by 2010 most had at least a third of their youth enrolled in colleges and universities, and seven had over one-half. At the same time, some of these, such as Pakistan, Kenya, and Nigeria, had yet to enroll beyond small proportions of youth in postsecondary schooling. And obviously some face significant political and economic challenges that will hinder advanced education and scientific research into the future. Nevertheless, there is significant room for postsecondary expansion in these countries.

Also, evidence of two additional factors indicates that university-based scientists in most of these countries are engaged in processes supporting mega-scientific research. First, research from these countries includes comparable numbers of high-quality studies. For example, Australian, Israeli, Chilean, and New Zealander scientists published 30 percent or more of their papers in journals with the highest quartile of impact (citation index),

Country	STEM+ Articles with Scientists from Country	% Increase in STEM+ 1990–2010	% with University Author	Enrollment Rates of Postsecondary Education in 2010†
Australia	30,674	200	85	80
Brazil	28,447	893	87	30
Turkey	19,897	2,375	90	56
Iran	16,336	13,862	95	43
Israel	9,430	95	69	66††
Mexico	8,190	560	64	27
Argentina	6,673	257	68	75
South Africa	5,888	122	90	19
New Zealand	5,659	163	76	83
Egypt	5,108	239	82	33
Pakistan	4,317	1,279	82	7
Chile	4,141	367	88	66
Saudi Arabia	3,164	252	84	37††
Tunisia	2,441	1,035	24	36
Colombia	2,217	1,440	89	39
Nigeria	2,110	142	92	10††
Algeria	1,513	1,073	82	29
Morocco	1,232	744	65	14
Venezuela	1,064	176	78	78
Jordan	965	408	93	40
Kenya	940	175	42	4
Bangladesh	934	624	72	10
Mean	7,334	1,204	77	40

Table 8.1. Countries Outside of Europe, U.S., and Asia with Most STEM+ Articles in 2010. Sources: SPHERE database; UNESCO Institute of Statistics.
† Proportion of Youth Attending (Gross Enrollment Rate): Lee and Lee (2016)
†† 2010 data from UNESCO Institute of Statistics (2021)

whereas the average among all twenty-two countries in 2010 is 18 percent. Second, as will be explored in the following chapter, like scientists worldwide, those in these countries are heavily involved in cross-border collaboration, with an average of almost one-half of annual papers including an author based at an organization in another country.

Like a large river resulting from scores of small tributaries, significant flows of STEM+ research from traditionally less-involved countries and regions is occurring—and these flows are growing. Both competition in scientific activity and spreading normalcy of the university-science model will continue to increase worldwide, especially spurred on by the massive expansion of education and science systems in East Asia and elsewhere. Indeed, China, over the past several decades, has taken an extraordinary globalization approach to education and science leading to collaboration of its university-based faculty with academic partners in other countries along the "New Silk Roads." With attempts to attract its promising scientists working in other countries to return, China is set to contribute to twenty-first-century science in the same way U.S. based faculty-scientists did in the twentieth century.[24] Yet even small states and larger ones not involved until late have shown their dedication to fund ambitious experiments in capacity building via university institutionalization. Thus the paradigm of university institutionalization spreads across all regions and to states of vastly different size. Whatever the future holds, these young or reoriented older universities are now the official national standard-bearers; simultaneously, they are internationally oriented, and with various versions of the university-science model are involved in all aspects of mega-science.

BRAZIL: CHALLENGES TO GLOBAL ADOPTION OF THE UNIVERSITY-SCIENCE MODEL

While universities have become increasingly similar across the world in their general knowledge-production missions and organization to achieve them, they all must exist in national environments of higher education funding and management. And some environments are more adaptable than others to postsecondary development and the drift towards the university-science model. In key ways, the continued globalization of mega-science hangs in the balance of what may happen to both the education revolution and the enactment of the university-science model in these new places. A good case from which to explore these challenges is Brazil.

By 2010, Brazil-based scientists, usually researching at one of its universities, were authoring just over one-half of all STEM+ articles that year from Latin America, and well above Mexico, the second most productive country in the region that year at 16 percent.[25] Also, since the mid-1970s Brazil increased its publication level at a rate similar to that of Taiwan and India, and with the largest population in the region, scientists in this country were the thirteenth biggest producer of publications in the world between 2011 and 2016.

Over the same period, Brazil expanded its set of universities and enrollments, but more slowly than in many other countries. Since colonial times, the expansion of opportunity for advanced education has been slow, and like other countries in Latin America, Brazil has struggled to cope with the demands of access to education, a problem that is accentuated in higher education. But in the late 1990s, through government investment and a privatization policy allowing growth in tuition-charging private universities, the system grew with noticeable increases in science capacity. In less than twenty years from 1998, the country increased its annual PhD production, including STEM+ degrees, from just under four thousand to over twenty thousand, with 70 percent of enrollment in private universities. And as shown above, by 2010 its enrollment rate in postsecondary education was into the rapid expansion phase.

Yet just beneath what appears to be the much-repeated expansion of postsecondary education and the science-university model are some limitations that could inhibit the country's continued participation in mega-science. Restraints loom from both within the university system and the country's politics. Even though overall expansion of higher education is occurring, it is still below the average of other economically developing countries, and others in the region, such as Argentina. The system also suffers from both an inefficient centralized bureaucracy and tension with localized authorities. Probably most problematic for the further development of a university-science model is that only about one-half of professors have a contract that includes the necessity of participation in research and publishing, plus faculty hiring at the public universities is more of a civil servant process than searching for scholarly and scientific talent.

Recent political corruption and instability in Brazil led to a populist, illiberal regime resulting in less than competent management of the economy and recovery from the COVID-19 pandemic. As will be discussed more fully in the Conclusion, a rise in illiberal ideologies and nationalist political parties are often at odds with both the expansion of higher education and development of research capacity. While much attention is given to shrill antiscience pronouncements from some countries' leaders, the retarding of advanced education has the most potential to limit enactment of a university-science model and the spread of global mega-science. Along with a few other countries, Brazil illustrates how the model is vulnerable to factors beyond the control of the postsecondary system, although they may be mostly extreme ones. At the same time though, there is no evidence among the newcomer countries that some revival model for marshalling societal resources for science is emerging or even discussed. And the world's science engine, backed by the education revolution, is now well established with its dynamic international collaborative networks growing stronger, a phenomenon examined next.

CHAPTER 9

Chasing Neutrinos Through Networks of Science

University Collaborations and Scientization

Tall and wiry with unruly curly black hair, physics professor Doug Cowen has expressive eyes that flash wide with contagious enthusiasm for explaining the mysteries of particle physics. With skill honed by years of creatively teaching undergraduates the physical essence of the universe, Cowen spins an engrossing tale about colleagues' and his attempts to trace the ubiquitous but most elusive of the universe's fundamental particles—the neutrino. It is a tale of major global collaboration, without which this amazing scientific search would not have been possible, as is increasingly the case for much of STEM+ research and universities.

Along with the better-known subatomic particles quarks, photons, and electrons, neutrinos played a major role in the formation of the universe. Indeed, without them our sun's fusion furnace would not function, and life could not exist. The existence of neutrinos was first suggested by the Italian physicist Enrico Fermi, who spent time at the University of Göttingen in the 1920s and developed the first nuclear reactor at the University of Chicago after World War II. As physics progressed, it became increasingly clear that learning about this class of particles offered valuable new clues for solving mysteries about everything from the Big Bang to possibly detecting them from deep space to test astrophysical theories. Yet despite their status as the most abundant massive particle, studying neutrinos is a challenge. Sometimes called the "ghost particle," these sand grains of the universe are neutral in charge and so tiny that they continually pass through objects, from our bodies to whole planets, mostly without leaving a trace. For example, some of an estimated one hundred trillion neutrinos propelled into space as relic particles of the Big Bang are imperceptibly traveling through the average-sized room every second. Neutrinos also have the scientifically useful qualities of not being influenced as they travel in mostly straight lines from their release points. And unlike other particles of matter, they can escape from dense regions surrounding objects like supermassive black holes, whose energy is occasionally released like an arrow in a jet of matter traveling at the speed of light, known as a blazar. It was known that once released, when neutrinos infrequently interact with matter it should result in puffs of light which could reveal their existence. Finding these puffs, however, requires an elaborate effort.

To attempt to find evidence of neutrinos and where particular ones might have originated from, Professor Cowen is part of a super-collaborative project known as the IceCube Neutrino Collaboration. More than three hundred astrophysicists, physicists, and engineers from fifty-two institutions, mostly universities, across a dozen countries are intensively coordinated to chase evidence of neutrinos.[1] At the heart of the collaboration is a laboratory, a small dorm, and a basic airplane runway in Antarctica, the Amundsen-Scott South Pole Station.[2] During the austral summer from November to February, when temperatures are a balmy −10 to −60 degrees Celsius, a few dozen hardy scientists and technicians work there. The IceCube Neutrino Observatory was built in this inhospitable and unlikely environment precisely because of the South Pole's deep, clear ice. Below the snowpack, across about a cubic kilometer of solid ice, Cowen and his colleagues have melted eighty-six holes and lowered over five thousand very sensitive light detectors. Neutrinos, of course, blithely travel through these hard sheets of ice, occasionally giving off a tiny amount of light, while other particles from the continual cosmic-ray showers from the Earth's atmosphere mostly do not. So in the ice these other better-known, or "just boring particles" according to Cowen, can be mostly ignored, and attention can be paid primarily to the signatures left by neutrinos. Rare as they are, these neutrino light detections can be recorded by the photomultiplier tubes embedded in the ice, turned into analyzable data, shared among the large collaborative team, and published in scores of journal articles, often with over three hundred collaborating coauthors, and rightfully so.

Reaching well beyond the capacity of any one lab or university, it is the second phase of the neutrinos project that brilliantly illustrates the scientific value of large-scale global collaboration. It is one thing to find evidence of neutrinos, it is a far harder task to find where in space any particular particle started its unstoppable journey to Earth. Accurately determining the origins of particles is essential information for the testing of astrophysical theories. Meticulously tracked through high-powered observatories, a kind of endless movie of the history of the universe unfolds from light and other electromagnetic waves released by violent astrophysical events millions of years ago, such as exploding stars, gamma-ray bursts, black holes, and neutron stars. "Because neutrinos can travel for billions of light-years through space without being deflected or absorbed, [if traced] they can uniquely provide accurate information about the distant universe, particularly the tau neutrino, which at high energies carries the undeniable pedigree of distant astrophysical origin," explains Cowen. With these observations of the past informing new astrophysical theory, predictions of where in the vast universe to watch for new evidence of these past events adds to the historical account. Neutrinos can add a major new chapter, but given their elusive nature, it is hard to know where to look for them, and hence to learn more about the telling cataclysmic events that sent them on straight lines towards us.

If Cowen and his international colleagues could detect neutrinos when they passed through the ice of the South Pole and immediately relay a reasonably precise direction of where they came from, there was a chance that some telescope-equipped satellites and multiple high-energy observatories around the world might be able to "see" and pinpoint the source. But this would take perfectly coordinated collaboration among NASA's (U.S. National Aeronautics and Space Administration's) satellites, a worldwide network of observatories, and the IceCube Neutrino Observatory. A unique multistage complex collaboration would have to occur in real time following a detection of neutrinos from an event in the ice, including communication of approximate, but accurate enough, directional origin in space, rapid shifting of satellites, and sufficient tracking by observatories across the globe. Coordinated by IceCube collaborators and aided by a computer relay from Cowen's part of the team at Penn State University, a high energy neutrino was detected in the ice in late 2017, and coordinates of its line of origin from space were then relayed in minutes from the South Pole to orbiting satellites and ground-based telescopes, several of which then found evidence of the neutrino-releasing blazar coming from four billion light-years from Earth—a major scientific success made possible by collaboration, or as Cowen proudly exclaimed, "teamwork wins the day."[3] And the movie of the universe's shaping continues, with the IceCube team continuing to detect hundreds of neutrinos that originated outside the solar system.[4] Furthermore, data from this long-standing, complexly managed, multiple-investigator, extremely expensive project is shared with scientists worldwide for more analysis, including machine learning using all available data, and publications. And so, the collaborative cycle begins anew.

A FLOOD OF COLLABORATION

As impressive as the size of the IceCube Collaboration is, in this respect it is just one of many large-scale collaborations across different sciences in which a large, expensive instrument is used to run experiments and collect specific data for large networks of scientists across the world, mostly in universities, to do their cutting-edge research.[5] Other high-energy astronomical messengers, particle physics instruments, gigantic telescopes, and the mapping of the human genome are similar super collaborative projects.[6] And these special collaborations lead to whole new scientific sub-areas, reflected in the expansion of the STEM+ topics underlying the dramatic growth in reports of new discoveries.

The average collaboration may be neither as massive nor as spectacular as the IceCube project and other super collaborations, but working in teams of scientists across vast distances and many time zones is now thoroughly routine. The once-familiar image of the scientist plodding along, alone in a laboratory, if ever true, is now far off the mark. Collaboration of various kinds spreads with mega-science, and in turn deepens the dimensions of future science. And this is not just science done with more and more assistants. Research and scientific publication have departed the age of scientific nationalism and entered an era in which collaboratively produced research across national and cultural boundaries is increasingly prominent. Such international, intercultural teamwork facilitated the unprecedented exponential growth in scientific knowledge at least since the 1980s, and the university-science model is at the heart of this.

Forms of collaboration beyond one scientist, or, in many cases, even beyond the small team, increasingly drive every aspect of the scientific process: theory development and conceptualization; methodological design and operationalization; gathering, sharing, and interpreting data; and finally publishing results after peer review for communication with ever more interconnected communities of investigators. This process, which is repeated over and over, builds capacity upon which scientific inquiry expands well beyond what would have been the case if most research were done less collaboratively. While science is often portrayed as a steady accretion of very small packages of new knowledge punctuated by sporadic, singular breakthrough discoveries and theory, a maturing pattern of collaboration makes intentionally planned, regular boosts in knowledge from large-scale investigations a normal feature. Breakthroughs cannot be fully anticipated. But concerted, well-planned, and consistently executed collaborations speed both accretion of findings and innovations with the potential for expanding new and established areas of inquiry as never before.

As shown in the case of the IceCube, collaborative scaffolding of research produces data open to new analyses by large numbers of scientists who were not necessarily involved in the data collection at the South Pole nor the tracking at a telescope. Such sharing maximizes the potential for new ideas, critique, and discoveries that enable new understanding and paths for future research. Whether it is data from the collective effort to map the human genome, massive amounts of astrophysical data, or the shared use of sophisticated instruments across many teams, working together to solve similar problems expands science. Even the results from prior studies on a particular topic are now turned into data for other scientists' meta-analyses to reveal new statistical patterns not observable via any one study. A collaborative environment also raises the standards of research so that working in teams becomes more than an option. In most fields of science today, the work is so complex that individual scientists cannot achieve meaningful results without collaborating—the so-called "collaboration imperative."[7] Indeed, this has become the established way to conduct research, particularly for those aspiring to conduct the most cutting-edge and influential research. If formal evaluations, performance measures, and continuously generated comparative indicators reflect a learning race to achieve new knowledge, participation in networks and interorganizational linkages with continuous communication and collaborations of different sorts is increasingly crucial to success.[8] Interculturally collaborative and internationally comparative research projects are more complex because such teamwork is demanding and the principles of effective research designs are more difficult to achieve.[9] But with extensive collaborations, often through networks linking universities and scientific fields, becoming the norm, younger scientists are socialized to master these processes by being exposed to extended professional networks and experience of senior faculty scientists providing the connections needed to foster innovation and provide synthesis.

What is further effective about networks of scientists is that smaller subteams or individuals can bring deep specialized knowledge of technique, design, context, and analysis to bear on a wider range of scientific topics by working together with others. This in turn opens many possibilities for the transfer of methods and findings from one sub-area to another. In a sense, these migrate via boundary-spanning collaboration into other sub-areas, often opening new perspectives for research. Although organizationally useful in the university, boundaries among traditional subjects, such as physics, chemistry, or biology, are somewhat arbitrary and come with intellectual costs that collaboration can mitigate by bending boundaries and creating useful hybrids of topics.

Although science has always had the potential to be collaborative, science by teams and networks of teams in multiple locations has exploded in quantity since the 1980s, especially due to information and communication technologies. The most visible, reliable, and accurate (but conservative) indicator of multiple scientists doing research together are the resulting coauthored journal articles.[10] By 2010, over a fourth of the world's papers across all fields resulted from scientists affiliated with research organizations in at least two different countries, or "international research collaboration."[11] Much more prevalent are collaborations in the same country and organization. As the solid line in Figure 9.1 shows, coauthored papers already made up almost three-quarters of the world's approximately 370,000 STEM+ publications appearing in 1980. A decade later, over 80 percent of publications resulted from collaboration, and by 2010 over 90 percent of the over one million annual articles across STEM+

fields were coauthored. As already shown in the case of Germany (Chapter 7), the intensity of collaboration has increased too: in 1980, the average collaborative paper was coauthored by two to three scientists, with most collaborations including no more than five coauthors. But by 2010, this number had grown to six coauthors on average with most collaborations falling in the much wider range of seven to thirty coauthors. Also, as demonstrated by the IceCube case,[12] that same year there were several hundred super-collaborative publications, including one hundred to over three thousand coauthors.[13] Such complex "hyper-authorship" is becoming increasingly common.[14] Indeed, most medical research, including mitochondrial vaccines so crucial in the fight against COVID-19 and a huge share of contemporary biological science, is based on the unraveling of the human genome, a collaborative resource initially involving the coordinated efforts among several thousand scientists across multiple countries. Nevertheless, the vast majority of international coauthored publications remain collaborations between two or three countries, but all coauthored worked is cited more.[15]

COLLABORATIVELY TRANSFORMED SCIENCE

Some increase in coauthoring is no doubt because of changing norms of receiving credit for involvement in science and a widening circle of roles in research that are now routinely credited on papers. At universities, this is particularly true, as postdoctorates, graduate students, and now even undergraduates wish to be involved in research and strive to receive publication recognition. No doubt, the ever-increasing volume of papers reinforces a rising standard of publication productivity for faculty and those training with them, so coauthoring is an advantage in meeting climbing expectations—and a crucial dimension of competition for reputation. But it would be too cynical to assume that these changes drive the majority of collaborations. If anything, it is likely the other way around: collaboration makes itself imperative. Teamwork, once made increasingly accessible, expands science to more topics, raising its potential for greater discovery and faster tests of theory. Although collaboration alone is not sufficient to produce high-impact research, it is increasingly necessary, not least due to the complexity of many research designs requiring diverse skills and

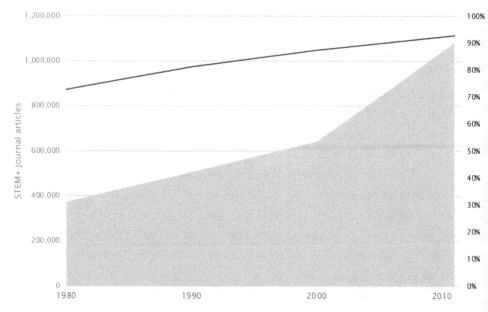

Figure 9.1. Increasing Collaboration Reflected in Proportion of Coauthored Publications Among World's STEM+ Journal Articles, 1980–2011. Source: SPHERE database.

the advantage in accessing a wider potential audience for one's findings.[16] In other words, widespread collaboration enhances all dimensions of scientization, and to have a chance at being influential, scientists have little choice but to collaborate, often in very complex networks, powered by information and communication technologies. Not that collaboration is without challenges and costs, but the motivations and learning processes for successful teamwork have grown, becoming standard practice across most fields.

Another of these dimensions is, as seen before, the rising globalization of science. Not only have more countries paired the development of higher education with university-based research, but greater collaboration across universities has flowed into the globalization of the research process itself. Not surprisingly, domestic collaboration led the way prior to this. For selected years, Figure 9.2 shows all coauthored papers by domestic and international coauthorships; in 1980, a year in which nearly all coauthored articles were from teams of scientists within the same

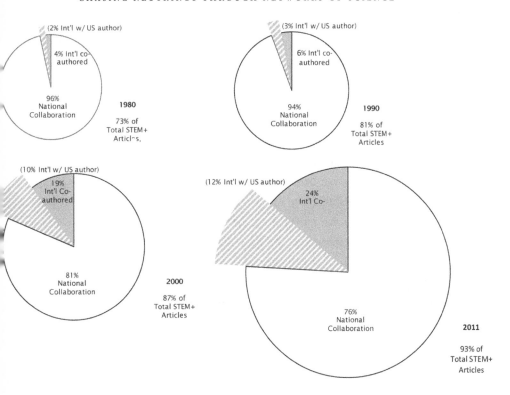

Figure 9.2. Coauthored STEM+ Journal Articles with Domestic Only or International Scientists, and Proportion of All International Articles Coauthored with U.S.-Based Scientists, 1980–2011. Source: SPHERE database.

country, only a small sliver included international collaboration. Yet over the ensuing decades, domestic collaboration's share steadily decreased as international collaboration increased. By 2011, most papers were coauthored, while international collaboration had grown to one-fourth of the approximately nine hundred thousand coauthored articles that year. And these internationally coauthored papers were overwhelmingly published by university-based scientists. At least since 1980, universities have produced more coauthored papers, both domestically and internationally, than other types of organizations. There was at least one university-based scientist on just over 60 percent of all domestic collaboration articles in 1980 and just over 80 percent by 2011.

University scientists were early to international coauthoring. By 1980,

84 percent of such papers included university-based coauthors, and this rose to 94 percent along with the significant increase in the amount of internationally collaborative articles. Not only did university scientists lead the way in collaboration, when scientists from universities are coauthors, scientists from nonuniversity organizations often also contribute. Of the international collaborative papers including a university-based author in 2011, two-thirds also include at least one nonuniversity coauthor. The constellation of organizational forms contributing most to cutting-edge publications in leading journals varies by discipline and location—as well as by the collaboration portfolios developed over time in particular organizations.[17] Collaboration and globalized research did not occur across all of STEM+ at a uniform pace; some disciplines, such as particle physics and astrophysics, were collaborative before others. Over the development of universities and mega-science, however, collaboration across all topics increased, first slowly and then exponentially, bolstered by information technology, education and scientific exchange, and extensive experience in international collaboration and the resulting networks. In short, universities have been the backbone of the collaboration driving so much of mega-science.

The inner workings of the IceCube Collaboration illustrate the power of networks of scientists driven by the university-science model. As described earlier, the organization of research in large universities means that Professor Cowen teaches and researches in an educationally subsidized college of science. Within one university, he helps fund and leads a neutrino research team made up of a research (nonteaching) professor, several postdoctoral scientists starting academic careers, three graduate students receiving advanced training, and several aspiring undergraduate astrophysicists. Cowen's home team is then embedded within a network of teams from forty-seven other universities across a dozen countries, each one within a similarly educationally subsidized college. This forms the foundation of collaboration upon which a smaller number of specialist nonuniversity organizations are added. It would be both scientifically inefficient and prohibitively expensive to form one stand-alone, three-hundred-scientist institute dedicated to research about neutrinos. What might have been the early-century approach under the Harnack principle guiding research institute formation has given way to the flexible and collaboratively accessible network of universities that host mostly autonomous researchers.

Practices of research production are also adapting to burgeoning collaboration. Cowen, for example, spends two to three hours a day in virtual meetings with co-investigators and other collaborators worldwide (even before the COVID-19 period). Aside from regular meetings about the maintenance of the detectors at the observatory in Antarctica, at any one time about twenty working groups focus on various aspects of the results, completing analyses, and writing papers. Their results are routinely posted online for the entire project team to see, critique, and review. There is also a fair amount of self-governance required for such a sizable project, so the project's Collaboration Board negotiates and enforces team practices and rules for proposing new papers, adding new collaborators, and managing recognition.

Extensive use of collaboration is also transforming age-old logic behind attribution of scholarly credit for ideas, research effort, and the publishing of articles. Some interpret the growing size of teams and numbers of coauthorships as trivializing credit or perhaps as a corruption of the intellectual process. With the hypercollaborative papers as evidence, some science observers, not to mention humanists, who are themselves on a slower yet inevitable path towards more collaboration, express dismay over the authenticity of contributions to sizable coauthorships. While such a critique makes a point, it misses the major implications of the bigger process. Of course, large groups of thirty or more scientists cannot jointly do analysis and write a paper in lock-step; such tasks are always done by smaller groups within the larger collaborative. In a sense then, these hypercollaborative papers do trample on traditional norms of authorship attribution and scholarly recognition. But such papers also represent a new form of scientific communication emerging from in-depth collaboration, in which credit is based on participation in very large networks of scientists with diverse strengths to contribute to the overall production of knowledge.

To date, Cowen's network has published approximately a hundred papers with nearly all three-hundred-plus IceCube collaborators as coauthors.[18] Along with this project's clear rules for "opting-in and opting-out" of any one paper, checks on free-riding, and guards against exploitation of early-career colleagues, the logic is as simple as it is necessary to motivate all involved. If one is a participating member of the collaboration, then they receive credit for the science that emerges. As this logic spreads, it has the potential to change older practices of attributing credit solely through articles. Indeed, Cowen notes that increasingly in astrophysics, in-depth evaluations by other scientists of a scientist's intellectual contribution are already gaining precedence along with article counts and citation measures of scientific impact.

Where exactly the flood of collaboration will take attribution of scientific credit and old notions of authoring and publishing is hard to predict. It is clear though that these large networks driven by university-based scientists are routine. Collaborative scientific work and publication will only increase and be more common as mega-science continues to unfold. How quickly this dimension proceeds will depend on how global scientific networks and funding agencies adapt and evolve to support such opportunities for discovery.

HUBS OF MEGA-SCIENCE

The extensive globalization of science has spread through collaborations with first American and then European universities, particularly since 1980. For selected years, Figure 9.2 shows all coauthored papers by domestic and international teams, and the raised proportion of international coauthorship including at least one U.S. scientist. For example, in 1980 three-quarters of the world's papers (370,000) are mostly coauthored by scientists in the same country, with just 4 percent produced by international collaboration, and over half of this includes a U.S. scientist. In 2011, while the U.S. share in this had dropped to 41 percent, it represented some hundred thousand STEM+ international articles authored in collaboration with U.S. scientists, the vast majority of whom were researching at the country's universities. In addition to working with U.S. researchers, scientists from other countries often coauthored with European university-based scientists, representing 47 percent and 42 percent of all international collaborative articles in 1980 and 2011, respectively (not shown in this figure). This also includes considerable transatlantic collaborations. At first glance, this might seem obvious, given the shared history of and the exchange relationships between American and European universities, and the deepening of the university-science model that they employ, have been so central to the volume of scientific articles since the mid-twentieth century; thus there was simply more opportunity for collaboration with the rest of the world. This is true, but it is more complex than that. A case can be made that the university-science model, driven by a growing education revolution of access to universities, originating in Europe, elaborated by American universities, and then appearing again in European universities after World War II, formed two major, sustained hubs for scientific collaboration. Both of these would have a significant influence on the timing and landscape of increasing research productivity in Asian countries.

To illustrate this further, Figure 9.3 shows simplified network representations of article coauthorship between U.S.-based scientists and scientists in the countries in which most of the former's coauthors are based. The thickness of the link indicates the frequency of article publication between U.S.-based coauthors with the other country (all other collaborations without a U.S. author are omitted), and the size of circles indicates that country's overall production level, measured in the volume of papers in the SPHERE database. In 1980, approximately one-half of all international collaborations with a U.S.-based author (approximately thirty-five hundred articles) were with collaborators based in one of twenty countries. And six of these countries—Canada, the U.K., Italy, West Germany, France, and Japan—made up the lion's share of U.S. international collaboration and had universities already achieving higher levels of paper productivity. Over the next three decades, American universities—as the world's main science hub—developed increasingly dense networks, as reflected in a higher volume between the U.S. and its top collaborating countries. Also, the network shifted to include scientists in more countries from more regions, with perhaps fewer geopolitical barriers.

As shown in Figure 9.4, like American universities, those in France, Germany, Italy, and the U.K. also became hubs of global collaboration, not least due to the ideational, normative, and regulative support for cross-national collaborations in research and teaching motivated by the European Commission and other supranational initiatives aimed at

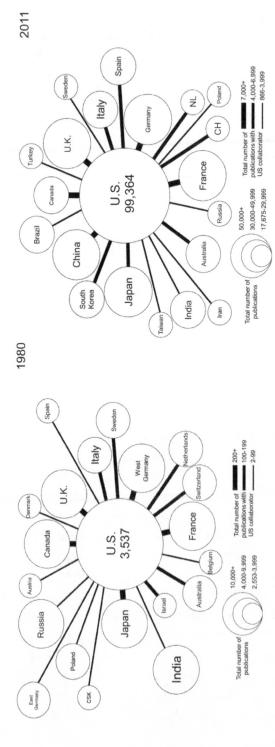

Figure 9.3. Network Representation of Scientists in the United States Coauthoring with Scientists in the Twenty Most Frequent Collaboration Countries, 2011. Source: SPHERE database. Note: Simplified network diagram; all other coauthored articles omitted..

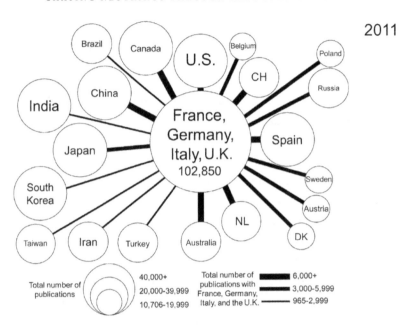

Figure 9.4. Network Representation of Scientists in the Four Most Productive European Countries Coauthoring with Scientists in the Twenty Most Frequent Collaboration Countries, 2011. Source: SPHERE database. Note: Simplified network diagram; all other coauthored articles omitted.

universities.[19] Mostly university-based scientists in these four countries together coauthored approximately the same total number of articles internationally as did their American counterparts, and by and large with the same countries. By 2011, the European region was made up of four major science powerhouses and numerous further countries with highly collaborative scientists, often based in some of the oldest and most well-connected universities worldwide.

Currently, American and European universities can be considered super hubs of global collaboration. Not only is there considerable joint research between scientists in North America and Europe, but scientists in many other countries have become involved. In 1980, U.S.-based scientists and those based in the four top-producing European countries were collaborating with scientists in about 100 other countries. By 2011, this had expanded to scientists in 193 countries for both hubs—a truly global research effort, inclusively reaching researchers in almost all countries. And the networks become denser as the volume of collaborative papers between countries increases. Consider, for example, the supposed competitors China and South Korea. By 1990, as these two countries entered into mega-science, university-based Chinese and Korean scientists collaborated with either Americans or Europeans in the big four on about 650 STEM+ articles. This pattern has continuously and markedly increased, such that by 2011, Chinese and South Korean scientists collaborated with U.S. or European scientists on about 23,500 and 7,800 articles, respectively. The flexibility and relative openness of the university-science model has meant that collaboration, first domestic and then international, has been a key factor in moving more university scientists from more countries into networks of discovery.

Existing networks tend to pull in more universities and their scientists. For example, the prestigious Korean SKY universities and two others—Korea Advanced Institute of Science and Technology and Kyungpook National University—published about six hundred collaborative articles with scientists at American universities in 2000. And in that year scientists from just five American universities—University of Illinois, Harvard University, Massachusetts Institute of Technology, University of California Los Angeles, and Northwestern University—were the main partners for all Korean-U.S. collaboration. Eleven years later, the Korean-U.S. network had become much denser, with scientists from a wide range of universities in both countries having copublished over seven thousand articles.[20]

The internationalized STEM+ disciplines and the strong regional hubs for collaboration of North America and Europe, mostly led by universities, has meant tremendous and rapid growth in collaboration, even when measured by a most conservative indicator: coauthored publications in leading, mostly Western (and English-language) scientific journals. Spatially, the number of countries with its scientists collaborating internationally has increased significantly within the span of a few decades. Most recent analyses, by Dag Aksnes and Gunnar Sivertsen, of fifty million WoS publication records across all disciplines show that while the share of internationally coauthored publications continues to incrementally grow—from 4.7 percent (1980) to 25.7 percent (2021)—the proportions continue to vary substantially across countries, from 30–90 percent, due to size of country and economic development.[21] Further factors giving rise to these persistent disparities are the resources invested; the variable institution-alization of higher education and science systems we have analyzed here; the differential impact of internationalization; and cultural, linguistic, and geographical differences, with disciplinary specialization reflecting these factors.[22] Universities everywhere now host scientists who increasingly and successfully collaborate across such borders. Global infrastructures, such as the IceCube, are necessary to provide crucial custom-made platforms for cutting-edge collaboration and to facilitate the processes of scientific discovery that transcend national borders to utilize the extraordinary net-works of contemporary science built and maintained by research universi-ties, and the other research-producing organizations, throughout the world.

CHAPTER 10

Conclusion

Global Mega-Science, Universities, and Their Joint Future

Never free of humanity, science is a social institution. Even as it progress-
es towards a highly technical unraveling of the unknown into strands of
facts and abstract theories few of us fully comprehend, it must navigate
the social world. This is a way of referring to a "culture of science" as
it adjudicates and disperses its rules, norms, and ethics, identifying what
of this immense undertaking is celebrated and what is hidden backstage.
As with any social institution, to be successful, a culture of science must
generate enough legitimation to recruit extensive and varied resources.
These resources are necessary to support the expensive, demanding, and
sometimes professionally risky activities of research. Seen this way, what
our journey through time and space reveals is unmistakable: *the legitimation and resourcing of global mega-science rest upon a more than one-
hundred-year mingling of a culture of science with a culture of higher
education.* Evolving ideas pushed the (higher) education revolution, just
as other evolving ideas behind science pushed the pace of research, and
in mixing these, each social institution lent the other all-important social
legitimation, and, increasingly, essential and shared resources. The fact
that higher education is ubiquitous around the world and mega-science
is now extensive and inclusive testifies to the longevity, durability, and
joint success of their complementarity within this cultural arrangement.
Universities have become ever more salient in the culture of education and
in societies more generally. They have been, and continue to be, the main
stage upon which this intersecting cultural dynamic unfolds. Universities
create a substantial and sustained global academic platform upon which
so much science builds—an essential space for the exchange of ideas, network creation, and concrete collaboration.

Nobel economist George J. Stigler once quipped that "intellectuals
are not inexpensive and . . . we professors are much more beholden to
Henry Ford than to the foundation which bears his name and spreads his

assets."[1] But while scientists, like the rest of university faculty, are indeed expensive, the explosion in their numbers and their research production over the century of science was not just a function of economic development. It took much more than simple capitalist expansion, with or without philanthropic spinoffs, to facilitate global mega-science. Had science not been powered by the deepening (higher) education revolution, it is doubtful that even the most robust private research and development programs and government funding imaginable could have come anywhere close to achieving what has, in fact, developed.

Nevertheless, single-factor theories of the origin of sweeping historical change are always tricky to advance. Behind long-evolving, complex phenomena there are many factors acting in concert, intended or not, and it is always possible for observers to be led astray by too much focus on any one of them. Yet just retreating to what is essentially an "everything causes everything" argument is as misleading as it is unenlightening. At its best, a single-factor argument focuses on what might be sufficient versus necessary, or what is the most crucial factor, among others, without which the change could not have fully occurred. As noted earlier, the seemingly inexorable rise of global mega-science has been facilitated by myriad factors that also grew over the century of science, such as national science policies and funding; geopolitical forces causing armament and space races; overall economic development and accruing wealth; technological breakthroughs; the dynamic of scientific discovery itself; and pressing societal crises such as famine, epidemics, and now climate crisis. Each has been considered extensively in other historical accounts and all have been shown to be influential at different points. But even if all were lumped together in one giant causal force, they would not have been sufficient to pull off mega-science. This is not to say that mega-science does not depend on other factors, but rather that it took something more to create the phenomenon at a global scale that shows no signs of slowing. Especially as the world grappled with the COVID-19 pandemic, we saw how important scientists—working collaboratively across the globe—are to solving recurrent threats to humanity.

Our journey emphasizes a central cultural process that provided an organized and well-resourced platform for undertaking unprecedented amounts of scientific discovery. The education revolution instilled an ideology that advanced formal education should no longer be limited to producing a small elite. Instead, first it became a course to develop more individuals than only a privileged few. Then, steadily in a growing number of countries, advanced education became an avenue for most individuals to reach their full potential and was equated with the path to achieve the good society. Whether always true or not, this ideology steadily spread. Higher education, increasingly believed in, everywhere, provides a powerful cultural partner for the dynamic activities of science and research. The reverse is also true: being a place of significant scientific research lends considerable prestige, legitimation, and funds to universities and other postsecondary organizations. The institutions of higher education and science have profoundly complemented and transformed each other dynamically over decades and centuries: these complementary institutions have become thoroughly mutually dependent.

Consequently, we have traced the ideas and their enactments in the forming of a university-science model across time and space. As noted, the term "university-science model" is shorthand for the result of progressive, long-term shifts in notions about the missions, organization, and products of both universities and the doing of science more generally. The historical evolution of ideas can be hard to track, as no one consistently names them or meticulously records changes as they progress. This is left to scholars to sort out long after the fact. Unlike concrete historical events and their enactors, ideas remain abstract and their consequences are often opaque. To make some sense of the usual historical messiness around transformative notions and their implementations, we employed the idea of a model. It emphasizes how pieces of this institutional relationship evolved over time and across events, often occurring in many different places without a lot of explicit intentional sharing or overt copying, and despite some misunderstandings about its driving forces. An image of an evolving university-science model as an overarching, yet loose, guiding form is more accurate than images of prescriptions, design blueprints, or centralized plans of omniscient science stakeholders. When an idea, such as that of the mission of a university to centrally include scientific research, is supported by teaching each new generation of scientists, it gains widespread significant legitimation. The ensuing model of how things should be done to make them that way takes on a normal, taken-for-granted quality over time. This sets the overall framework that shapes detailed planning and concrete organization. Ideals continually shape actual decisions and concrete outcomes, often without awareness of the model influencing the local enactors. As challenging as it may be, follow the ideas behind a robust model and the chaos of history recedes. In the case of global megascience, deep commitments to higher education and the doing of science begin to make more sense.

Along with expert training, universities have always focused on knowledge and scholarship, yet the domains of knowledge—and efforts to generate it—were limited over much of academia's early history. With time, ever more domains were integrated. Scientific discoveries forced their way in, making the curricula of the late-nineteenth-century university considerably more robust. It became the leading edge of an epistemological shift, first in the West and then globally.[2] Social theorists have long noted that the privileged ways of knowing and operating in modern human society have shifted towards those that take on qualities of rationality. And this overarching paradigm includes science applied to ever more domains and things. The university's knowledge generation and teaching of a wide array of subjects have become major rationalizing agents of society. This is the observation at the center of Talcott Parsons's insightful extension of the earliest sociological thinking about rationality and postindustrial trends, motivating his coining of the term "education revolution" and theorizing its importance as part of the foundation upon which liberal society, including capitalism and democracy, is built. Aggregate consequences of the university-science model make his prediction salient for understanding a range of significant recent phenomena, such as the rise of the knowledge society, changing economies, growing inequalities between old jobs and newer ones requiring advanced education, and, of course, the relentless scientization of society. Mega-science, through the development of the university-science model, is exhibit number one for any trial of Parsons's original observation, now shared by other sociologists, of the encompassing impact of the educational revolution on society.

Nothing cultural is truly unique, but the eventual migration of actual research from the earliest European specimen clubs to academies of science and ultimately to thousands of universities and their faculties marks a clear point of departure. As academies hatched a more systematic and robust culture of science, the university became its nursery. It evolved as a place explicitly merging teaching and research. What has become known as the "Humboldtian" vision, as a coalescing of ideas about the university and research, guided this development. As this occurred, the first stirrings of the advanced ideology of the education revolution led towards ever-greater enrollments in upper-secondary education. This would eventually bring more youth from differing backgrounds into university training, across all disciplines in preparation for all kinds of occupations. There, connections with social status, new types of jobs, and occupational placement, as well as the beginnings of the idea of cultivation of a larger segment of society, were consolidated for the first time in history. Later, this evolved into the sweeping contemporary image of mass personnel and human development through more and higher education, including lifelong learning and occupationally specific continuing education. Combining extensive research and a Humboldtian ideology with a broadening legitimation of advanced education, this new university-science model was hardly in full focus by the late nineteenth century. Even so, the historical record of an uptick in the pace of discovery is clear. Already evident at the University of Göttingen and a handful of other similarly research-intensive universities in Europe, these supporting forces created a dynamic between greater institutionalization of education and scientific inquiry that was already ushering in scientific activity at an unprecedented level. This would be the key cultural pattern enacted across different regions of the world over the next century; repeated and innovated upon in an accelerated and intensified fashion, even to the farthest areas and smallest of countries, as we have seen.

This budding brew of cultural forces pulled the world's center of scientific gravity to the eastern coast of North America between the World Wars. Two decades into the twentieth century, several dozen universities in the U.S. joined twice as many in Germany and other European countries in contributing to a steadily rising rate of annual STEM+ papers with an accompanying rising rate of those authored by university-based scientists.

Personified by Evan Pugh and C. N. Yang, the journey's cultural analyses of additional developments of the model reveal processes that would repeat themselves many times over in the U.S. A country fueled by a maturing postsecondary education revolution in the thirty-year period after World War II significantly increased the number of universities and generated unheard of capacity for scientific research. The U.S.'s haphazard development of higher education is marked by comparatively more access for a broader and increasingly diverse population of students.[3] This is reflected in massive enrollments from across American society, including hundreds of thousands of international students in graduate education. Today, this global recruitment and training produce the most important raw material of science and source of innovation—young scientists. Crucially, access has been facilitated by all kinds of founders of colleges and universities, including a commitment by postsecondary organizations themselves to engage in some degree of self-direction and the morphing of their charters to open their doors ever wider. From a range of founding scenarios, these multiple factors combined to form a self-reinforcing synergy.

The result is visible throughout postsecondary education in the dynamic expansion of research universities, such that STEM+ training and research spread horizontally across smaller elite universities to larger but less-prestigious public ones, as well as everything in between. Eventually, the university-science model saturated vertically down to include all types of other postsecondary schools, including those not previously known for research contributions, such as the growing number of two-year colleges, formerly mainly vocational and oriented towards applied sciences. Extending well beyond the most prestigious universities aimed at elites, this U.S. version of the university-science model became the backbone for the maturing trends of mega-science worldwide.

A near infatuation with the admittedly spectacular science produced by the world's leading universities in the U.S. obfuscates the true scale of the full expansion that the university-science model created. Similarly, scholarship has overemphasized the early explicit emulation of the German university by a handful of late-nineteenth-century American universities. In fact, some organizational elements, such as reserving the university for only graduate training, were misunderstood. Such an arrangement, if fully enacted, would have missed out on the advantages of the significant cross-subsidizing from undergraduate education that exists today in the world's most differentiated higher education system. Instead, the real story behind the U.S.'s massive contribution to mega-science is the now-taken-for-granted university-science model that enabled hundreds of thousands of scientists to undertake STEM+ research at hundreds and then thousands of postsecondary organizations of various kinds, found in every corner of the country.[4] The most research-active higher education organizations provide positions for faculty-scientists training new PhDs and postdoctorates to push the cutting edge. All the while, they produce ever more, and arguably also more sophisticated, research. These papers are published in thousands of peer-reviewed journals oriented to global scientific communities now collaborating to enhance both the quantity and quality of scientific discoveries.

The same cultural forces, but rapidly condensed, allowed the grand entrance of some extraordinarily productive East Asian countries into burgeoning global mega-science. These cases illustrate an expeditious diffusion and demonstrate how the university-science model was transferable across cultural boundaries and funding sources. To a degree, the model has been malleable in the face of other cultural traditions of higher education and science, including contrary organizational forms that carry out the most advanced research, from academies and research institutes to hospitals and firms. Too often the historical consequences of ideologies are assumed to occur only by imposition, say from colonial, hegemonic actions or direct borrowing of organizational forms and establishment of specific organizations. While there were influences, such as the post–World War II "reform" of Japanese education, along with explicitly American ideals, the university-science model was attractive enough to be adapted to fit within existing arrangements in many contexts from different starting points. Interplay among a rapidly unfolding education revolution, national desire to participate in mega-science, and a persistent Mandarin ideology made for contrasting pathways to essentially the same outcome as in much of Europe and North America. In addition, even with the largess of excellence initiatives, the unexpected successful dissemination of the university-science model well beyond the chosen few universities in these countries is further evidence of the attraction and adaptability of the model to all kinds of higher education systems.

The sturdiness of the model is further demonstrated by the scientific success of universities in post–World War II Germany, despite their weakened legitimation and receding resource allocation as higher education expansion has exceeded all predictions.[5] A culture of genius, flowing in part from earlier notions of the elite professorship within the expansive nineteenth-century university, led to the independent research institute and a partially enacted dual-pillar policy justification emphasizing training over research within the country's universities, relatively similar in status due to their common emphasis on research. To the larger point, over the course of mega-science there have been numerous alternative suggestions in many countries that somehow the best-and-brightest scientists should be identified and provided with (far) greater access to high levels of resources versus open competition for proposed projects among all scientists. The underlying assumption is that this would naturally lead to greater amounts of peak science. This assumption has hardly been rigorously tested, even though it regularly attracts interest among science policy experts.[6] Prestigious German research institutes, particularly those of the Max Planck Society with their guiding Harnack principle, are perhaps the closest to a fully enacted and funded best-and-brightest model in the world today. Yet there is not much evidence that these institutes, in the aggregate, significantly outperform research universities. The former certainly contribute to pushing the boundaries of knowledge in specific subfields, and their scientists are often well-networked globally. But without the cross-subsidy of providing educational services and the wider cultural support of the education revolution, stand-alone research institutes are an expensive societal proposition. Even with less-than-optimal funding and societal legitimation, universities in Germany remain at the very core of this country's highly prolific science system.

This is not to conclude that the independent institute is an unsuccessful organizational form. These institutes assemble tremendous collections of leading scientists, including numerous Nobel Prize winners.[7] In the U.S., much crucial research is performed at nonuniversity centers, from the National Institutes of Health to Bell Labs (now Nokia Bell Labs), plus hybrid forms, such as the Rockefeller Institute (now University), that have contributed crucially to scientific breakthroughs.[8] But across the entire expanse of science, no other platform of research has succeeded in generating enough cultural and material support to outperform the university-science model. Plus, as seen in the organization of contemporary large American research universities, the university-science model includes some similar advantages of a kind of research institute reproduced many times over in academic departments and their colleges (faculties).

Most important, all research organizations and their scientists are increasingly embedded in a continuous competition as well as a global conversation and effort to discover the next breakthrough—via collaboration. This truly worldwide collaboration, witnessed even on the ice of Antarctica, requires an institutional scaffolding and organizations that accept permeable boundaries. The core organizational form of the research university, open via its multiple missions of research, teaching, and public service, provides the central platform for researchers in other organizational forms and around the world to join forces to address their fields' most pressing problems. Yet until the century of science, only a handful of countries had the vision—and the resources—to invest in this model substantially and sufficiently.

Returning to Europe via the Middle East, we saw how two very different leaders, in disparate national contexts, both gravitated towards a similar vision. In Qatar, an otherwise enlightened woman in the anachronistic role of an absolutist Arabian queen established a platform for international higher education and scientific innovation according to global norms. In Luxembourg, at the heart of Europe, a gender-barrier-breaking, entrepreneurial pianist-turned-politician founded a millennial national flagship research university. Both women facilitated the transfer and adaptation of the university-science model to small but wealthy countries attempting to leap forward in borderless competition at the nexus of higher education and science. The melding of advanced graduate education with scientific investigation matching global standards in the research-active university has truly become an accessible, taken-for-granted, and, to a degree, flexible model for the new century. In each case, there was little need for commissions that traveled the world in search of unique ideas to form their strategy; the way forward was obvious for them. What Qatar and Luxembourg ultimately illustrate is how by the end of the twentieth century widely accepted variations on the university-science model had become globally diffused and fully legitimate: even the smallest countries aspired to and indeed successfully created the environments and exchange platforms for scientific mobility necessary for research universities, becoming part of the infrastructure of global mega-science.

The further globalizing of the model is evident in the growing number of countries that have made up the steady flow of science from regions not originally very active in research. The accretion from just over two dozen countries at mid-century to over two hundred countries with scientists, mostly university-based, regularly contributing papers to leading journals by 2010 is remarkable. While the many yet-to-be-documented stories from these countries will no doubt present inevitable twists, turns, and setbacks, it is likely that at their core most will have implemented some form of the university-science model.

The world is tilting towards greater production and parity in production of scientific research because of the continued global impact of the education revolution. The ideas driving this were not borne out of a few single events as were many other periodical boosts to research on specific topics. Similarly, although some make out distinct stages of science, small to big, we instead see the drivers of mega-science there from the start, gradually but consistently intensifying, more than forming qualitatively different phases over the century of science. There is no need for an overly complex theory of shifting origins by periods within the century of science; better instead are the self-reinforcing dynamics of this exponential growth.

The journey also makes clear that mega-science and the university-science model are at the center of ongoing scientization, and, more generally, the insistent rationalization of society. The latter has been a major theme in understanding the difference between postindustrial society and earlier ones. Adding in the evidence here of a dynamic but ultimately self-reinforcing relationship between the education revolution and STEM+ research, the epitome of a rationalizing process, strengthens this argument. This conclusion parallels and enhances evidence of other forms of scientization of the polity and the state worldwide.[9]

The accompanying analysis of the constant flow of millions of STEM+ studies also reveals an explosion of collaboration, embedded within a systematic, consistent, and interconnected world of scientific research. As noted, the frequently invoked media images of "Chinese," "American," or "German" science and national winners or losers of an international competition of ratings and rankings are hardly accurate anymore, if they ever really were. The wide enactment of a university-science model makes for a familiar, manageable, and common organizational environment for the world's hundreds of thousands of active scientists—in upwards of thirty-eight thousand universities. Across the networks, scientists, projects, and results are mobile, virtually and often physically too, and this is the essence of what is meant by the globalization of scientific inquiry, ever more inclusive. Certainly, restricted commercial R&D development and patents as well as secret military technological development are still the domains of companies and national governments. But at its base, the science involved is usually built upon the open, accessible record of discoveries in the public domain. And in truth, very rarely are new findings kept secret for long before they become swept into the linked processes of mega-science, powerfully exemplified in the origins of the Internet itself.

A bird's-eye view of what occurred since 1900 worldwide in the deepening of science as an institution with the university as its main platform is summarized in Table 10.1.[10] As our journey illustrates, when postsecondary education emerged and developed so did the major characteristics of global mega-science. STEM+ paper publications, numbers of countries with contributing scientists, scientization of new topics reflected in new journals, and all types of collaboration, including international, robustly grew. At the start of the twentieth century, less than one percent of the world's youth attended some form of postsecondary education, and even by mid-century when European countries, Canada, the U.S., and a few others had sizable enrollments, the world rate was still tiny. Expansion since then has been substantial, so that by 2010 just under a third of the world's youth attended higher education. Although the percentages before 1980 might appear trivial, they represent one of the largest recorded changes in modern society. Recall that in 1900 most of the world's population was completely unschooled, and for a country to enroll even a modest number of postsecondary students it must first develop primary and then secondary education for many children. Providing basic education for most usually takes several generations and sustained massive government support for each level to blossom and ensure wider enrollments. This must also include the establishment of universities and other postsecondary organizations, an additional large societal undertaking that mostly occurs after advanced secondary education is established for a significant portion of youth. Until the middle of the century, most countries in the world likely had one national university and maybe a few others, but this changed as development of earlier levels of schooling produced ever more potential students for postsecondary education. The most recent estimate indicates that in 2022 just over 40 percent of the world's youth were enrolled in some form of postsecondary education.[11]

Extensive educational development and changing social norms have played out in many nations. And as we have shown, one of the major consequences of this is the increase in universities and other postsecondary institutions with faculty-scientists contributing research through various enactments of the university-science model. From the data on STEM+ papers, we estimate that in 1900 scientists in approximately 130 organizations published at least one paper. It took the next fifty years for this to double, even though the capacity for more research grew much faster as the capacity at many universities increased. And concurrent with world enrollments, by 1980 this had exploded to over seven thousand universities and grew to a staggering thirty-eight thousand forty years later. As we have seen, not all postsecondary organizations produce similar amounts of research; some such American two-year colleges publish only a handful of papers each year while the hundreds of faculty-scientists at any large research university routinely publish thousands of papers collectively each year. Regardless of this variation, the point is that universities and other postsecondary organizations are the home of mega-science, continuously increasing capacity for research.

Last, the account here presents a fundamental challenge to the perennial intellectual debate over whether universities and science are in decline.[12] Excessively fatalistic views of scientific growth and arguments that the university has lost its power as an organizational form always

Dimension	1900	1950	1980	2010	2020
Volume of Science Production (# STEM+ papers, rounded)	10,000	48,000	373,000	1,200,000	3,000,000
Globalization of Science (# Countries producing leading journal papers)	26	36	162	206	
Scientization (# STEM+ journals)	92	486	3,902	8,337	9,526
Collaboration % of papers coauthored	75	78	80	92	
% International collaboration	0.3	1	2	22	
Universities and other Postsecondary Organizations Publishing STEM+ Papers*	130	260	7,270	38,200	
Worldwide Enrollment Rate (% Youth in Postsecondary Education)**	0.2	1.3	12.4	29	40

Table 10.1. Worldwide Dimensions of Mega-Science and Expansion of Postsecondary Education, 1900–2020.Sources: SPHERE database; World Bank. *Estimates only from countries with discernable postsecondary education in 1900 and 1950; estimates from all countries reporting enrollment rates in 1980 and 2010 (data: worldbank.org/indicator/SE.TER.ENRR). **Estimated from data on STEM+ publications data, rounded; considerable disciplinary differences exist.

attract attention. Yet, ironically, these dire predictions often create motivation for even more science. More seriously, they rarely, if ever, come to terms with the actual trends of mega-science; they miss the fundamental connection to the education revolution and all that implies. This history does, however, raise the question: How sustainable is this extraordinary, globe-spanning university-science model, and with what implications for the future of global mega-science?

The Future of the University-Science Model and Global Mega-Science

Upon reading an early draft of the first chapter of this book addressing the dimensions of mega-science, a colleague exclaimed, "There has to be an upper limit to this madness!" Madness or not, global mega-science continues. Whether mega-science has become some runaway collective mistake is ultimately a question of the value of science. As in the past, economic and societal cost-benefit questions will be heatedly debated. In any case, we have seen that mega-science continues its exponential growth. An estimate of the total volume of new STEM+ publications in 2019 found that the world's leading scientists published more than three million new journal articles documenting their newest discoveries, and, of course, most of these journals had university-based scientists contributing, reviewing, and editing. For minor technical reasons, this estimate includes a somewhat wider variety of publications than the earlier annual ones analyzed here, but it indicates that the mega-science growth curve continues—one million in 2010, two million in 2015, and about three million just four years later. While the doubling time of the flow might slow and wars and climate crises may dampen overall production for the next few years, the key dimensions of scientific research and publication patterns appear to remain very robust. While huge, the global system of research and collaboration can be nimble and effective, as witnessed by the profusion of research relating to COVID-19 and the successful global vaccine development.

Considering science's future in light of what we now know about the influence of the education revolution adds an otherwise missing perspective to the question of sustainability. If anything is clear from the record of papers published over this very long time frame, it is that predicting exactly when mega-science will taper off is a fool's errand. The persistent underappreciation of the scientific engine—driven by the education revolution and densely networked universities—proved those wrong who have posited predictions about the decline of science in the past. And those who have recently predicted a new imminent ending may soon join them for the same reason. Even though de Solla Price, from his early vantage point, missed the coming globalization of the academic platform for science, he was nevertheless correct to note that exploding trends eventually level off to slower rates of growth. But how quickly that happens depends on the sustainability of the necessary resources behind the trends—in this case the ultimate carrying capacity for world science. That is why even though we could have stopped the journey at tracking the many newcomer countries and regions into global mega-science, we proceeded further to a last stop in Antarctica with a focus on the collaborative dimension of research, which has once again dramatically increased the potentials for scientific discovery across boundaries.

Today, there is no essential "national" science, but rather unbridled collaborations for new knowledge, most often implemented with data and tools unheard of even a few decades ago. National and international science policy institutions and initiatives increasingly promote the creation of large consortia of universities to advance scientific agendas and solve problems more efficiently (for example, the EU's Horizon 2020 program).[13] Likewise, countries are more likely to employ science as part of their diplomatic strategy as a kind of "science diplomacy" of joint funding programs to promote collaborations among higher education and research organizations across borders.[14] With constant communication and decentralized efforts to break new ground, across cultures and organizational boundaries, the costs of collaboration, which are substantial, decline, and the benefits are ever more evident, for individual scientists and their organizations.[15] Global mega-science is today being driven as much by collaboration as by competition among scientists.[16]

The consequences of this point to several ways by which mega-science could sustain itself and even generate the potential for more growth. One obvious fact is the sharing of resources and their strategic global allocation. This is demonstrated by the massive IceCube Project's sharing of both expensive physical infrastructure of the unique hardware and software of the study and the global intellectual resources in the combined talent of the investigators. This is truly a swarm of intelligence and research experience—and so different from the lone-genius model of Harnack's stand-alone, pioneering research institutes. Shared research infrastructures from talent swarms to physical instruments to pooled massive datasets make the doing of research ever more collaborative, massively expanding the capacity for yet more discovery.

Alongside creating economies of scale, an additional, yet less appreciated, consequence is how collaboration creates new areas of research—an essential, self-reinforcing dynamic driving further scientization. Detailed histories of large collaborative projects often include accounts of how unexpected, or hard-to-achieve, findings can yield whole new empirical approaches and the conceptualization of new areas for more research.[17] Universities are particularly well suited for the sponsoring of this expansive process as they are the main organizational form responsible for the intergenerational transfer and transdisciplinary exchange of knowledge. Yet the knowledge-creation activities of all types of universities are not fully appreciated, nor are the long-term benefits of investments in this infrastructure routinely calculated.

Beyond promoting collaboration, universities bring even more unique sustaining influences. After a set of new discoveries leads to a new area for research, invariably at some university enterprising faculty venture to form a new graduate program with associated curricula and degrees. Some of these do not last, but others do and spread, in some cases very rapidly, across universities worldwide. The training of scientists dedicated to a new subfield follows, enabling further conceptualization and legitimating research activity within it. The cultural power to give definition and boundaries to new areas of knowledge, reinforced by degrees in specific curricular areas, makes universities a dynamic force in growing scientization—and its ultimate sustainability. No other type of organization involved in research—not national labs and private companies nor national and private research funding bodies—can wield such fundamental influence on science. Of course, new societal problems, technology races, national agenda, and the like also shape new avenues for research. The bringing of years of relatively unheralded genetic research on viruses, much of it done at universities, to bear on vaccine development took unique coordination beyond what universities can do. Government agencies, such as the NIH in the U.S. and pharmaceutical firms working with researchers in diverse organizational forms have been essential; indeed, the networks of such interorganizational relationships proved an essential source of innovation.[18] But, with its curricular control and degree-forming charter, only the university can so authoritatively demarcate and validate new sub-areas opening up for research—and immediately begin educating the next generations of scientists to carry forward research. This occurs so seamlessly within universities that its uniqueness is considered natural and routine, but it is remarkable and goes to the heart of the broader impact of the university-science model. When combined with its more open collaborative, multidisciplinary environment, this considerably bolsters this model's sustainability.

Aligned with this is the ability of universities to be open to new scientific pursuits, including the unestablished, risky ones. Motivation for risk-taking in university-based research does not derive from some bottom line of profit, nor ultimate social utility or technological apps. Nor does this rely solely on external funding, which is only one, limited, means to more research, and is often less supportive of pioneering, risk-taking research than standard "normal science."[19] Instead, the driver is mostly curiosity and intrinsic motivation for new discoveries, and the approving communal judgment of other scientists. As deflating and difficult it can at times be to achieve, it comes as close to an endogenous sustaining process of science as ever existed. While universities are rigidly administrative and bureaucratic as all like-sized complex organizations, the sacred notion (sometimes and in some places challenged) of academic freedom to investigate within them manages to create a crucial degree of flexibility in trying out new areas for research. As noted at the start of the journey, the Humboldtian ideology was always more about intellectual freedom than political expression, and while there have been numerous attempts to limit or control inquiry over the course of the century of science, mostly these have failed, as in Göttingen.

Since universities are central to these dynamics, the question of sustainability and growth turns on the future of the dynamics underlying the university-science model. There are reasons to predict its continued positive impact on science. At the same time, there are some trends indicating that the model may have run its most expansive course. For two sets of countries, Table 10.2 displays the number of STEM+ papers with an authoring scientist per 100,000 people in the country in 2010, and the now-familiar enrollment rate of youth in postsecondary education.

The first set of countries are those from the first chapter known to have scientists authoring high proportions of the world's total papers, and the second are the twenty-two most productive newcomer countries (or already productive but from regions less associated with earlier growth) described in Chapter 8.[20] Both indicators are broad and thus a bit simple, but telling nonetheless. Several things become clear. When differences of population sizes of countries are taken out of the story, some have gotten more out of their institutionalization of science—mostly university-based, along with some other productive organizational forms—than others. Switzerland, Sweden, and the Netherlands all publish considerably more papers per population than the average among all the traditionally high-performing countries. This also indicates that, all else equal, there could be less room for significantly more research capacity in these places in the future. If, however, these higher-per-population levels are looked at as potential upper bounds of what may be possible for other countries to reach in the future, they suggest significantly untapped capacity for research left in the world. For example, very large countries, such as China and India, had considerable room for capacity growth as of 2010. And in fact, recent estimates indicate that at least China continued to expand its capacity, such that by 2019, in absolute number of total papers, Chinese university-based scientists have surpassed scientists based in American universities.[21] Except New Zealand and Israel, both of which have incorporated their versions of the university-science model for some time, many of the newcomer countries also likely have more potential capacity.

The inverse of the postsecondary enrollment rates, a kind of measure of how many youth are still to be included in the cross-subsidizing of research via their education, tells a similar story. As of 2010, most of the newcomer countries had a significant share of youth still to be incorporated into postsecondary education. Of course, it is hard to predict both educational and research capacity development across a range of countries, some of which have slowed or reversed overall movement towards a liberal society for a range of reasons. But even if less than all of them progress as have others, in the near future there is likely a possibility for sustained growth of the university-science model and mega-science. Certainly, the world will not any time soon get to the point that every young

Traditional High Publication Countries	Papers per 100,000 People	Gross Enrollment Rates in Postsecondary Education[†]	Newcomer High Publication Countries	Papers per 100,000 People	Gross Enrollment Rates in Postsecondary Education[†]
Switzerland	235	53	N. Zealand	130	83
Sweden	177	75	Israel	124	66[††]
Netherlands	146	64	Turkey	27	56
Australia	139	80	Chile	24	66
Canada	129	69	Tunisia	23	36
U.K.	111	61	Iran	22	43
Taiwan	93	87	Argentina	16	75
Germany	91	57	Brazil	15	30
U.S.	86	93	Jordan	14	40
France	86	56	S. Africa	12	19
Spain	80	78	Saudi Arabia	12	37[††]
S. Korea	74	100	Mexico	7	27
Italy	72	64	Egypt	6	33
Japan	51	58	Colombia	5	39
Poland	43	74	Algeria	4	29
Russia	17	75	Morocco	4	14
China	10	23	Venezuela	4	78
India	3	18	Pakistan	2	7
			Kenya	2	4
			Nigeria	1	10[††]
			Bangladesh	1	10
Mean	84	63	Mean	22	38

Table 10.2. Scientific Discovery: STEM+ Journal Articles per 100,000 People, and Higher Education Development in Traditional and Newcomer High-Performing Countries, 2010. Source: SPHERE database.
[†] Lee and Lee (2016)
[††] UNESCO Institute of Statistics (2021)

person will attend university; nevertheless, expansion continues apace. And while the usual predictions of overeducation and runaway credentialing will be put forth as limits to the spread of the university-science model, there is scant evidence that this is based on an education bubble waiting to burst. Economies and other institutions are being transformed by advancing education more than ever before without convincing evidence of educational inflation.[22]

It is certain that even without sizable increases in rates of growth in capacity, the current global one will continue to churn out mega-science levels of discovery every year. This is driven especially by scientists in ever more countries participating in the extraordinary growth of *collaborative* research, which further diffuses cutting-edge knowledge across scientific communities that are no longer necessarily geographically proximate. Collaboration and increasing levels of education provide opportunities for growth and contributions that were unthinkable just three decades ago.

Running counter to these optimistic predictions are the known threats to further development of advanced education: not enough people, not enough money, not enough good governance. The past record of education expansion indicates an amazingly robust process, but it is entering new territory, literally. The regions of the world with the largest cohorts of youth still to be included in advanced education are made up of countries that have been able to make significant, but far from perfect, progress in developing primary and lower-secondary education for most of their population. It is yet to be seen though, if many countries in sub-Saharan Africa, parts of Southeast Asia, and South America can turn past educational progress into postsecondary development, and to what degree. Many of the world's countries have continued the development of upper-secondary and postsecondary education, including some in these regions too. Yet because of past massive exploitation of natural resources and people and continued forms of world inequalities, these places are most vulnerable to all challenges to future economic, social, health, and educational development.

Although some regions still have significantly more youth to incorporate into advanced education, the regions that are traditionally high in producing STEM+ papers are entering into sharp fertility declines. South Korea

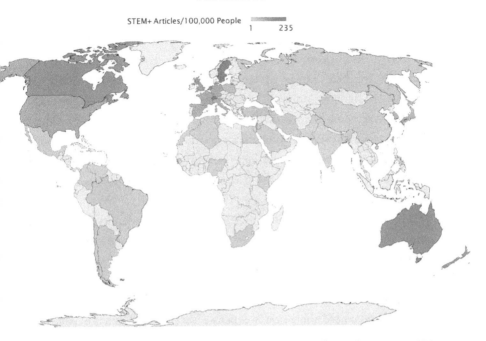

Figure 10.1. Mapping Scientific Discovery: STEM+ Journal Articles per 100,000 People, 2010. Source: SPHERE database.

and Taiwan, with the lowest birth rates worldwide, are a kind of "canary in the coal mine" of imploding future cohorts of university students and the closing or merging of universities. As shown earlier, both countries rapidly embraced a form of the university-science model and made great strides with their production of science, but now are seeing the effects of below-replacement birth rates that are rapidly shrinking cohorts of children. Closing and merging schools from primary to university levels, these countries' ministries of education are wrestling with maintaining a highly advanced educational infrastructure for a declining population. Demographic evidence has long established that most, if not all, of the world's population is headed to significant fertility declines, and with it the most basic ingredient for processes described here—waves of youth—also declines. It will take time, of course, for the full effect of this to be felt everywhere in education and science capacity. Even South Korea and Taiwan, with university mergers and closures, continue to keep pace and still manage to increase their annual level of paper production, at least as of 2019. And universities in many countries are becoming highly motivated to develop new educational services for lifelong learning (and credentialing) that will keep adults involved and voting for their share of government education budgets. But clearly the implementation of the university-science model will have to be significantly adapted to receding cohorts of youth in some of the currently most productive countries. At the same time, many traditionally high-producing countries continue to achieve extraordinary science production with state-funded universities embedded in education and social policy provisions—and being increasingly well-networked.

The future of the university-science model and advanced education could be disrupted financially as well, especially among those countries with more privatized higher education systems. This is clear from the long-term effects of rising tuition costs for individuals, and rising costs for governments in countries with long-established postsecondary education. Mass advanced education is expensive, even though it has masterfully cross-subsidized science. For numerous reasons, the cost of postsecondary education continues to rise—with long-term consequences. In a heavily privatized system, such as those of the U.S. and the U.K., rising costs pose a significant barrier to a larger proportion of families sending their children to colleges and universities. Smaller colleges and universities increasingly find themselves in financially precarious territory; some must close or merge. Similarly, although education is usually one of the largest government investments in most countries, adding a greatly expanded advanced education system takes revenue demands to a whole new level. Thus, in meeting rising demand for postsecondary education among families beyond what traditional state-funded higher education could provide, many countries, such as Peru, Ecuador, and some in Eastern Europe, have had negative experiences with runaway for-profit universities of dubious educational and research value. Off-setting this negative is an effort to improve their quality in the future, as various countries aim to require research activities also in these universities. Also, some increasingly avail themselves of supranational programs, such as in the European Union, and a number of these will survive and thrive by moving in that direction.

Decisions around costs and funding could be further problematic for sustaining mega-science to the degree that national governments continue to organize themselves as if education and STEM+ research were only trivially connected domains, activities, and budget demands. Unmindful, or only superficially aware, of what has been examined here, most countries maintain separate ministries for education and for science with all the usual intergovernmental policy and funding rivalries and lack of coordination that this implies. Governance via ministries and other agencies, not only through increasing R&D investments, leads them to seek to increase their influence on research agendas through the proliferation of specifically defined thematic programs.[23] In the U.S., for example, science and education policy and guidance comes from a dizzyingly complex array of agencies, including the National Science Foundation, National Institutes for Health, and the U.S. Department of Education, as well as fifty state agencies for (higher) education, R&D, and innovation. As touched on in discussion of such cases as Germany and South Korea, regardless of the degree of centralization, a separation of higher education and science governance can be problematic, although sometimes it does offer some flexibility for new approaches. It is important for the future that the world's education and science policymakers more fully recognize how intertwined and complementary these two institutions are. Innovative policy with full appreciation for how much universities and other postsecondary organizations are the necessary platforms for science, and in turn how scientific research helps to legitimize and extend the relevance of universities, will be essential for countries to maintain and even expand their participation in—increasingly collaborative—research, eventually helping to solve the wicked problems the world faces. Even newer entrants to global mega-science, such as Luxembourg and Qatar, can quickly reap the societal and economic benefits of investment in universities, that, founded in the collaborative era, have from the very start been embedded less in nascent national higher education and science environments than in the globe-spanning networked ones.

The model, however, is not invincible, and a final set of challenges centers around politics and governance. Various political threats to liberal society in recent years, such as Hungary's currently claimed "illiberal democracy," include, if not an outright anti-education ideology, then a distinct reluctance to support particular universities or even the wider freedom of inquiry that is necessary for scientific advance.[24] The same is true for these regimes' resistance to politically inconvenient science. To the degree that populist, illiberal political forces continue in the world, they will be a counterforce to sustaining a widespread university-science model. Also, governments are always tempted to harness research capacity in their countries to meet specific national goals. For example, in recent years Qatar's government has limited and narrowed along more nationalist interests the range of funding for research at its Education City research conglomerate.

Likewise, education development does not fare well in failing states. For example, as of 2010, Venezuela, as one of the newcomer countries, was making strides in growing postsecondary enrollments, the university-science model, and participation in mega-science. Yet its more recent combined political, economic, and social crises have reduced these capabilities and have set them back for the foreseeable future. Even China, the country with the now largest absolute output of STEM+ research papers, with considerable potential for more in the future, has placed certain limits on the autonomy of its scientists, leading to frustration about less creativity ascribed or fewer citations given to this research.[25]

Returning to the question of what "too much" science would be, it is easy to assume that the dimensions of mega-science, propelled as they have been by the university-science model, could be excessive. But that assumes a narrow social utility to all things. Over a century of sociological, anthropological, and general social science research points to the fact that humans create meaning ascribed to their activities as much as they wed them to (immediate) utility, the latter of which is never fully obvious anyway. This is partially Parsons's point as to why the university has come to play such an important role in promoting the idea that rationalized and scientized knowledge is the most meaningful, now in almost every aspect. Science moves to address issues *and* to its own internal rhythms; looking to it for ultimate good or evil is in most cases misguided. It is highly doubtful then that mega-science will collapse under its own weight. It is an activity that holds too much meaning, along with its partner institution mass education. And thus attempts to engineer to some imagined optimal outcome, level, and distribution always fail because it is impossible to fully engineer culture.

Regardless of what the future holds, our journey reveals an amazingly expansive story, unexpected and still underappreciated. When our team began to collect the information on STEM+ papers and hone our hypothesis, it seemed a bit risky, because comprehensive estimates over the century of science were lacking to show where new discoveries came from.[26] Even the basic dimensions of mega-science were only partially known. Of course, it has been apparent that science continues to expand and that universities play a role in that, but we did not expect the evidence and the historical arc of this strengthening relationship to be so clear. The university-science model, ideologically charged and resourced by a world-sweeping, maturing education revolution, is the indispensable force behind mega-science.

The self-reinforcing growth of higher education and science has exceeded all expectations. But this surprise is due to the taken-for-grantedness, even misunderstanding, of the power of the university and the symbiotic relationship of higher education and science that has unfolded over centuries and in all regions of the world. These two cultures never were fully independent of each other. Universities have always proved to be society's most effective repositories of living knowledge, whatever the political conditions, and networked scholars and scientists across borders. And by fusing research and teaching, the research university is also the master of intergenerational knowledge transfer—and thus the cultural guide to an even more scientized future. The creative fusion of these cultures in the organizational form of the university strengthens both cultures, legitimating the investment in the capacity of the next generations to do more science and sometimes to solve myriad human-made crises through further discoveries. As we continue to face durable inequalities, novel pandemics, and fierce climate crisis, the collaboration at the heart of global mega-science—so vividly demonstrated by the collaborative efforts called forth in the global response to the COVID-19 pandemic—seems well-placed to provide our last best hope of a sustainable future.

Acknowledgments

As we have demonstrated, today collaboration is essential for science and among scientists, often working together from different points around the world. The same is true for this sociological study of the development of mega-science and its supporting universities and other research-producing organizations over the century of science. An international team of social scientists, working at universities across the time zones in America, Europe, the Middle East, and East Asia, collaborated for ten years to produce a huge longitudinal dataset on knowledge production, an extensive series of technical papers, several dissertations, and an edited volume collecting in-depth country cases—and finally this book synthesizing what we have learned. The historical journey described here rests on our expansive team's intellectual labor and teamwork across cultural and status group boundaries, and we are especially grateful for their innumerable contributions, diverse skills, and culturally specific inspiration. We thank the interviewed scientists who gave us insights into their fields, their mobilities, and the ways they pursue their careers in an era of collaboration.

This project on and by border-crossing research collaboration has relied, from the start, on international exchange programs (Fulbright, FNR Luxembourg) and resourceful organizations devoted to scientific advancement. We first met two decades ago in Berlin at the Max-Planck-Institut für Bildungsforschung (MPI Human Development), but the original, stimulating discussions exploring what would become "global mega-science" took place in the garden of the Wissenschaftszentrum Berlin für Sozialforschung (WZB Social Science Center Berlin) with the late Gero Lenhardt and Manfred Stock. In the decade since, the project has been university-based, at Penn State University and the University of Luxembourg, with the bridge sustained by David's multiyear guest professorship in uni.lu's Department of Social Sciences.

The SPHERE project received funding from the Qatar National Research

Fund, a member of Qatar Foundation (NPRP Grant No. 5-1021-5-159), to purchase the original data and recode it extensively. John Crist, then directing research at Georgetown University School of Foreign Service in Doha, got it all started and facilitated the project's intensive first phase. Doctoral candidates Jennifer Dusdal, Yuan Chih Fu, and Seung Wan Nam coded and constructed the database and delved deeply into country contexts for their dissertations; Wan Yu and Yahya Shamekhi conducted further analyses of the SPHERE dataset. In addition, Wei Bao, Junghee Choi, Frank Fernandez, Hyerim Kim, Iris Mihai, Robert D. Reisz (1964–2020), Kazunori Shima, Liang Sun, Mike Zapp, and Liang Zhang contributed to *The Century of Science* (2017).

Over the years, we have benefited from valuable feedback received during innumerable presentations and discussions, many of them virtual, especially during the pandemic. We thank the organizers and diverse participants for engaging with our arguments and for probing questions as well as the organizations for hosting us: American Sociological Association, American University, Centre for Global Higher Education (London/ Oxford), Comparative International Education Society, German Centre for Higher Education Research and Science Studies (DZHW), German Society for Higher Education Research, Ghent University, Hertie School of Governance, INCHER/Kassel University, Institute for Higher Education Research (HoF)/Martin Luther University Halle-Wittenberg, Leibniz Centre on Science and Society (LCSS)/University of Hanover, Leibniz Educational Research Network (LERN) of the WGL, Luxembourg Educational Research Association, Papal Catholic University of Peru, Peking University, Penn State University (Culture and Politics workshop), Stanford University, Texas Tech, and Tübingen University.

We greatly appreciate the press's reviewers and Roger L. Geiger and Kayla Kemp at Penn State for careful readings of the full manuscript and their constructive critiques. David Baker thanks Raf Vanderstraeten and the colleagues at Ghent University, Belgium, for awarding him the 2023 George Sarton Medal and Chair for the History of Science. Justin Powell thanks the uni.lu and Q-KNOW teams, including Jennifer Dusdal, Anna Kosmützky, Marcelo Marques, Achim Oberg, and Mike Zapp, for helpful comments. In autumn 2020, a sabbatical at Oxford University provided time to focus (during lockdown) on forms of collaboration.

We thank SUP editor-in-chief Kate Wahl for her commitment to this book, as with our previous SUP books, and to Marcela Maxfield and Dan LoPreto for guidance over this long journey.

Coming home, we are deeply grateful to our families on both sides of the Atlantic, especially to Mimi Schaub and to Bernhard Ebbinghaus, for their steadfast support as this book gradually evolved—and our understanding of mega-science became more global.

Notes

1. The 2023 Nobel Prize in Physiology or Medicine was awarded to Katalin Karikó and Drew Weissman for their identification of a crucial chemical tweak to messenger RNA that enabled the development of effective COVID-19 vaccines that have saved millions of lives.

2. Brainard, "Scientists Are Drowning in COVID-19 Papers." Indeed, the extraordinary volume of research papers published now demands AI-based information retrieval and natural language processing should scientists hope to find the most relevant work related to their own research. The COVID-19 Open Research Dataset (CORD-19) was designed to connect diverse expertise in the "machine learning community with biomedical domain experts and policymakers in the race to identify effective treatments and management policies for COVID-19" (Wang et al., "Cord-19"). From March 13, 2020, to June 2, 2022, the date the final version was released, the weekly updated version of the dataset had grown to index over one million relevant papers (https://github.com/allenai/cord19).

3. Perhaps the most comprehensive global source of data on COVID-19 was provided by a university: the Johns Hopkins Coronavirus Resource Center (https://coronavirus.jhu.edu/), developed by the Center for Systems Science and Engineering in the Department of Civil and Systems Engineering, collected worldwide data on cases, deaths, vaccines, testing, and demographics from January 22, 2020, to March 10, 2023. The database registered 676,609,955 total cases; 6,881,955 deaths; and 13,338,833,198 total vaccine doses administered.

4. See, e.g., Diamond, *Guns, Germs, and Steel.*

5. See Schofer and Meyer, "The Worldwide Expansion of Higher Education in the Twentieth Century"; Frank and Meyer, *The University and the Global Knowledge Economy.*

6. National Science Board, "Publications Output: U.S. and International Comparisons."

7. Godin and Gingras, "The Place of Universities in the System of Knowledge Production"; Powell and Dusdal, "Science Production in Germany, France, Belgium, and Luxembourg"; Dusdal et al., "University vs. Research Institute?"

8. Peter Weingart ("Growth, Differentiation, Expansion and Change of Identity—The Future of Science") recounts that the growth of science has often inspired "visions of doom and destiny" while early analyses of scientific growth suffered from simplifications, false assumptions, and limited imagination to correctly foresee how science would develop over the long run.

9. Sparking debate, recent studies show that despite the growth in papers, the proportion of papers that are disruptive or considered breakthroughs—as opposed to incremental papers—has declined (see Park, Leahey, and Funk, "Papers and Patents Are Becoming Less Disruptive Over Time"; Kozlov, "'Disruptive' Science Has Declined").

10. Mike Zapp ("Revisiting the Global Knowledge Economy") shows that almost all countries had striking increases in their knowledge personnel, with OECD countries more than tripling their R&D personnel since the 1980s and non-OECD countries more than doubling theirs since the mid-1990s; see also Powell and Snellman, "The Knowledge Economy."

11. "Zombie Research Haunts Academic Literature Long After Its Supposed Demise."

2

On scientific fraud, see Franzen, Rödder, and Weingart, "Fraud"; see also Retraction Watch, "Retraction Watch."

12. Usually, observations of a connection between education and science stop at noting that the former spreads scientific literacy and trains the talented few to become scientists. Required to sit through typical science courses in school, most people are more scientifically literate than were their grandparents or parents, and there are increasing opportunities to gain advanced scientific training. But this leaves scientization—the greater institutionalization of a broadening scope of scientific inquiry through expanding domains and a deepening of the scientific activity within existing disciplines and society (see Murakami, "Scientization of Science" and "Transformation of Science"; Drori et al., *Science in the Modern World Polity*; Drori and Meyer, "Scientization" and "Global Scientization")—far too narrow, trivializing what has occurred. This includes the ways in which science has expanded and differentiated as well as how scientific knowledge has become a source of legitimacy across societal sectors (Gauchat, "The Cultural Authority of Science"). Alternatively, what we will refer to in shorthand as the *university-science model* formed a very potent and self-reinforcing connection between mass education and science, manifested in the symbiotic relationship between the university and science, without which it is unlikely that mega-science could ever have spread successfully to all continents and nearly every country.

13. David Frank and Jay Gabler (*Reconstructing the University*) uncover the shifting patterns of development in universities around the world of curricular offerings, disciplinary fields, and types of knowledge. Postsecondary curricula are becoming increasingly internationalized and isomorphic (Zapp and Lerch, "Imagining the World"), especially with the increasingly dominant *lingua franca* of English and scientific dominance of the United States since World War II (Stevens, Miller-Idriss, and Shami, *Seeing the World*).

14. Abrutyn, "Toward a General Theory of Institutional Autonomy"; see also Drori et al., *Science in the Modern World Polity*.

15. Jason Owen-Smith (*Research Universities and the Public Good*) emphasizes that research universities not only create new knowledge but also anchor (scientific) communities as well as serve as hubs for evolving diverse societal activities, which enables their continuous contributions to problem-solving and innovation; see also Stevens, Armstrong, and Arum, "Sieve, Incubator, Temple, Hub."

16. In the U.S., the *Bayh–Dole Act* of 1980 changed the game for university patenting: the university share of U.S. patents grew exponentially for two decades through 1998 (Leydesdorff, Etzkowitz, and Kushnir, "Globalization and Growth of US University Patenting (2009)."

17. UNESCO reports (2022) that R&D expenditures have continued to increase over the past decade, underscoring the fundamental importance of science, technology, and innovation (STI), especially during the COVID-19 pandemic. Countries strive to meet the Sustainable Development Goals (SDGs) Agenda 2030 of encouraging innovation, substantially increasing their research workforce and their public and private R&D spending (monitored in SDG 9.5). Data for 155 countries and territories shows continued growth in global R&D investment, despite the pandemic challenges: the average annual growth rate over the last decade (2010–2020) was 4.7 percent (3 percent adjusted for inflation). Over the decade, the proportion of global GDP invested in R&D went up significantly, from 1.61 percent to 1.93 percent.

18. Rudolf Stichweh (*Wissenschaft, Universität, Professionen: Soziologische Analysen*, first and second editions) charts the differentiation of disciplines and professions and further emphasizes that the university is simultaneously a local and global organi-

zation, with further regional and national relevance, due to embeddedness in networks and the building of universal knowledge systems (Stichweh, "The University as a World Organization").

19. In his tome based on extensive interviews and management experience, William Kirby (*Empires of Ideas*) charts the evolution of the modern university from Germany to the U.S. to China today, comparing the organizations, their resource bases, and the political conditions in which they operate.

20. Dusdal and Powell, "Benefits, Motivations, and Challenges of International Collaborative Research."

21. Murakami, "Transformation of Science"; Tenopir and King, "The Growth of Journals Publishing."

22. Powell, et al., "Introduction: The Worldwide Triumph of the Research University and Globalizing Science"; Dong et al. ("A Century of Science") also emphasize the shift from individual to teamwork amid the globalization of scientific development, analyzed on the basis of Microsoft Academic Graph (MAG), a dataset with 100 million publications from 1800 to 2016; Okamura ("A Half-Century of Global Collaboration in Science and the 'Shrinking World'") uses OpenAlex, another large-scale bibliometrics platform, launched in 2022 to replace MAG and now including 239 million works, to replicate change in the countries collaborating and contributing most to fifteen natural science disciplines over the past half-century.

23. See, e.g., Weingart and Winterhager, *Die Vermessung der Forschung*; Ball and Tunger, *Bibliometrische Analysen*; Gingras, *Bibliometrics and Research Evaluation*.

24. See, e.g., Gingras, *Bibliometrics and Research Evaluation*; Wang and Barabási, *The Science of Science*.

25. Laudel, "What Do We Measure by Co-Authorships?"

26. Officially launched in 1964, the Science Citation Index was created by Eugene Garfield at the Institute for Scientific Information (ISI); see https://clarivate.com/the-institute-for-scientific-information/history-of-isi/. The more selective Science Citation Index Expanded now includes over 9,200 of the world's most impactful journals across 178 scientific disciplines. From 1900 up to today, the index catalogs more than 53 million records and 1.18 billion cited references.

27. The follow-up project, "Relational Quality: Developing Quality Through Collaborative Networks and Collaboration Portfolios" (Q-KNOW) extended the analysis for Germany, focused especially on coauthorship patterns across disciplines and organizational forms, through 2020 (see Dusdal, Oberg, and Powell, "Das Verhältnis zwischen Hochschule und Wissenschaft in Deutschland").

28. The three main databases are sufficiently stable in their coverage for in-depth cross-disciplinary comparisons (Harzing and Alakangas, "Google Scholar, Scopus and the Web of Science"); on the evolution of altmetrics and their use for measuring societal impacts of research, see Tahamtan and Bornmann, "Altmetrics and Societal Impact Measurements"; and more generally, see Krüger ("Quantification 2.0?") on the implications of the massive expansion of reliable bibliometric infrastructures via digital publishing and automated data processing for data production and assessment.

29. See, e.g., Godin and Gingras, "The Place of Universities in the System of Knowledge Production"; Adams, "Is the U.S. Losing Its Preeminence in Higher Education?"; Bornmann and Mutz, "Growth Rates of Modern Science"; Bornmann, Wagner, and Leydesdorff, "BRICS Countries and Scientific Excellence."

30. Adams, "Global Research Report: United Kingdom," p. 6.

31. See Powell et al., "Introduction: The Worldwide Triumph of the Research University and Globalizing Science," for assumptions and procedures on stratifying by language, sample sizes, and weighting, adding to the sampling frame in earlier years, and other details.

32. But see the historical analysis and discussion of the proliferation of universities in Frank and Meyer, *The University and the Global Knowledge Economy*, chapter 2, which is based on *Minerva Jahrbuch der gelehrten Welt*.

33. See Dusdal, *Welche Organisationsformen produzieren Wissenschaft?*, for the organizational form matrix applied to the data; Dusdal et al., "University vs. Research Institute?", for comparison of German research universities and research institutes.

34. See, e.g., Gauffriau et al., "Publication, Cooperation and Productivity Measures in Scientific Research" and Gauffriau et al., "Comparisons of Results of Publication Counting Using Different Methods"; for a critical overview of various metrics and mechanisms and their implications, see Wu et al. "Metrics and Mechanisms."

35. See Powell, Baker, and Fernandez, *The Century of Science*.

36. Sugimoto and Larivière, *Measuring Research*; Wang and Barabási, *The Science of Science*.

CHAPTER 1

1. Shapin, *The Scientific Revolution*.

2. Price's *Little Science, Big Science* was his best-known book, published in 1963 (republished in 1986, with a foreword by Robert K. Merton and Eugene Garfield, other pioneering figures). Two years later, Price held the first Science of Science Foundation lecture, "The Scientific Foundations of Science Policy," at London's Royal Institution to promote evidence-based science policymaking. He published articles based on his lectures, including "Networks of Scientific Papers," and "The Science of Science."

3. Price, *Little Science, Big Science* and *Little Science, Big Science. . . And Beyond*.

4. The periodization little, big, and mega-science are used heuristically to make sense of developments during the long "century of science," but these should not be taken as sharp demarcations.

5. Czaika and Orazbayev ("The Globalisation of Scientific Mobility, 1970–2014") show several supporting trends to this in their analysis of bibliometric data of global scientific mobility over the past four decades, including increased diversity of origin and destination countries; the continuous eastward shift of the center of gravity of scientific knowledge production; and increased average migration distances of scientists associated with enhanced integration of more peripheral knowledge-producing organizations into the global science system. As we will see in the case of neutrinos research (Chapter 9), the "nature of different types of research practices implies different spatial relations that in turn influence the motivations for and outcomes of academic mobility and collaboration" (Jöns, "Transnational Mobility and the Spaces of Knowledge Production," p. 111; see also Jöns, Meusburger, and Heffernan, *Mobilities of Knowledge*; Gui, Liu, and Du, "Globalization of Science and International Scientific Collaboration"). Across disciplines, international research collaboration leads to more influential, often-cited research, especially among scholars in different countries—a key argument for further globalizing the scientific enterprise and recognizing the brain circulation and intercultural teamwork that facilitates recognition and impact across scientific communities (see Sugimoto et al., "Scientists Have Most Impact When They're Free to Move").

6. Borders and names of countries changed over the twentieth century for many reasons, especially due to wars. Here, we will generally refer to countries by their current names.

7. Population size and economic power do not automatically translate into scientific wealth (May, "The Scientific Wealth of Nations"), as indeed small but well-governed countries, especially long-standing democracies, succeed best in gaining most from their investments in R&D (Allik, Lauk, and Realo, "Factors Predicting the Scientific Wealth of Nations."

8. As explained in the note on data, used throughout is the "whole count" bibliometric method in which a paper coauthored by scientists researching in different countries is added as one to each country's total count of papers. For a helpful synthesis on scientometrics and standards for the measurement of research, see Sugimoto and Larivière, *Measuring Research*. In WoS analyses from 1980–2021, Dag Aksnes and Gunnar Sivertsen ("Global Trends in International Research Collaboration") show that while internationally coauthored publications overall have gradually risen from under 5 percent to over 25 percent in four decades, this global total masks considerable disparities in developments and the extent of internationalization, especially relating to country size, income level, and geopolitics.

9. Bozeman and Boardman, "Assessing Research Collaboration Studies"; see also Wagner, *The Collaborative Era in Science*.

10. Kienast, "How Do Universities' Organizational Characteristics, Management Strategies, and Culture Influence Academic Research Collaboration?"

11. See Zapp, Marques, and Powell, *European Educational Research (Re)Constructed*; Marques, Zapp, and Powell, "Europeanizing Universities."

12. Kwiek, "What Large-Scale Publication and Citation Data Tell Us About International Research Collaboration in Europe."

13. Web of Science Group, "*Web of Science* Core Collection," https://mjl.clarivate.com/home.

14. In *The University and the Global Knowledge Economy,* David Frank and John Meyer emphasize the global university expansion, especially the encompassing inclusion of cultural content and the university's key role in public society.

15. Zhang, Powell, and Baker, "Exponential Growth and the Shifting Global Center."

CHAPTER 2

1. For research exploring additional causal processes, see Guston and Keniston, *The Fragile Contract*; Kennedy, *Globalizing Knowledge*; Leslie, *The Cold War and American Science*; Smith, *American Science Policy Since World War II*; and Stephan, *How Economics Shapes Science*.

2. The so-called "Mode 2" model of science is among the most popular visions of scientific development, often applied broadly to suggest changes in contemporary systems of higher education, science, and innovation, with the assumption that universities will no longer be as dominant as in a supposedly earlier Mode 1 (Gibbons et al., *The New Production of Knowledge*; Nowotny, Scott, and Gibbons, "'Mode 2' Revisited"). Even though this conceptualization to describe the changing structures, conditions, practices, and outcomes of scientific work has become influential, as will be shown here, it does not fit the trends of what has, and continues, to occur (see Hessels and van Lente, "Re-Thinking New Knowledge Production" or Zapp and Powell, "Moving Towards Mode 2?" for the case of educational research in Germany).

3. Earlier in the century, communities of scientists working on a subtopic were often

small enough that authors were sufficiently familiar with each other to not need to include their organizational affiliation on papers. Although we do not know for sure, omitted affiliations could have been higher among university-based authors than those at other places of science, and if so, our estimates might be lower than actual numbers from 1900–1920, after which missing affiliations on papers dropped rapidly as the inclusion of author's affiliation became a standard practice of most journals.

4. Recent scholarly works emphasize that higher education in the United States, in its extraordinary scope and diversity, did not develop according to a national master plan, but rather was organically supported by a range of communities, stakeholders, and policymakers at various levels (see Labaree, *A Perfect Mess*). Indeed, despite several key universities in the nation's capital city of Washington, DC, there is not one national university reigning supreme—in contrast to many other countries (see Meyer, *The Design of the University*).

5. Powell, *Barriers to Inclusion*; Richardson and Powell, *Comparing Special Education*.

6. Recent insightful analyses of education and the economy during this period in the U.S. are informative on understanding a maturing education revolution; see, e.g., Acemoglu and Autor, "Chapter 12—Skills, Tasks, and Technologies"; Acemoglu and Autor, "What Does Human Capital Do?"; and Goldin and Katz, *The Race Between Education and Technology*.

7. Parsons in the late 1960s was already thinking and writing about various implications of an education revolution, yet not as a main topic until the 1973 book (Parsons and Platt, *The American University*). In "The Making of Parsons's *The American University*," Raf Vanderstraeten discusses the place of this late work within the Parsonian oeuvre and the lack of its reception due to sustained critiques of his earlier work when he was the dominant social theorist in the U.S. The book was coauthored by Gerald M. Platt (1933–2015) and prepared in collaboration with Neil Smelser (1930–2017), who commented on the manuscript in the epilogue. After decades of service to sociology in Berkeley and beyond, Smelser delivered The Clark Kerr Lectures on the Role of Higher Education in Society in 2012, which appeared as *Dynamics of the Contemporary University: Growth, Accretion, and Conflict* (2013) in which he picks up on Parsons's themes, analyzing contemporary trends and challenges that have added to already highly complex structures of higher education and pressures the system faces, from retrenchment of public support without corresponding shifts in governance and accountability; commercialization in higher education and the expansion of private educational providers; the rise of information technology and distance learning; and the explosion of contingent and part-time faculty that has eroded academic security and freedom. See also Geiger, *American Higher Education Since World War II*, for historical analysis of higher education and other intellectuals' thoughts in the U.S. during this period. And recently, there has been a revival of Parsonian theory (see, e.g., Treviño, *Talcott Parsons Today*) and theorizing about the influence of education on society (see Mehta and Davies, *Education in a New Society*).

8. Baker, *The Schooled Society*.

9. Brint, "An Institutional Geography of Knowledge Exchange".

10. We simplify some of Parsons's complex descriptions of what became known as "structural functionalism," particularly in using the term "culture" in a broader sense than he did. For a more recent perspective on this, see Abrutyn and Turner, "The Old Institutionalism Meets the New Institutionalism."

11. See, e.g., Mann, *The Sources of Social Power*, 2012.

12. A new, and partially tongue-in-cheek, name for this cultural pattern is WEIRD for "Western, educated, industrialized, rich and democratic" societies. How such a society gen-

erates a new sense of what is real and how that makes people behave and think differently from the past is now the subject of some psychological research that aligns with Parsons's notion of differing cultural realities around similar collective functions. See, e.g., Henrich, Heine, and Norenzayan, "The Weirdest People in the World?"

13. Parsons's ideas reflected a wider intellectual discussion at the time of how to consider the educational trends described earlier, including everything from thinking of it as wasteful overeducation to truly transformative; see, e.g., David Riesman's *The Lonely Crowd: A Study of the American Changing Character.* Research on the intensification of education as an institution worldwide took off after Parsons; see, e.g., Fuller and Rubinson, *The Political Construction of Education.* Also, detailed historical accounts of changes in the university and their role in research provide important evidence for Parsons's prediction (e.g., Geiger, *Research and Relevant Knowledge*) and emerging research on the academization of the economy (e.g., Stock, "Hochschulexpansion und Akademisierung der Beschäftigung"). Baker (*The Schooled Society*) discusses how thinking and research about a schooled society has evolved.

14. See Baker, "Theoretische Zugänge zur Bildungsrevolution in der post-industriellen Gesellschaft"; and for the original economic research, see Goldin and Katz, "The Race Between Education and Technology."

15. For new theory based on some of Parsons's ideas, see Abrutyn, *Revisiting Institutionalism in Sociology: Putting the "Institution" Back in Institutional Analysis*; and for social classes dominated by education factors, see Baker et al., "Education: The Great Equalizer, Social Reproducer, or Legitimator of New Forms of Social Stratification?"

16. Studies of the relationship between science and applications report that most STEM journal articles on new discoveries precede patents (see Narin, Hamilton, and Olivastro, "The Increasing Linkage Between US Technology and Public Science"; Meyer, "Does Science Push Technology?"; Geiger and Sá, *Tapping the Riches of Science*; Wible, "Patents from Papers Both Basic and Applied").

17. On scientization of society, see Drori et al., *Science in the Modern World Polity*; and for an overview of science and earlier societies, see Huff, *The Rise of Early Modern Science.*

18. A recent review symposium (Hoelscher and Schubert, "Universities Between Inter- and Renationalization") contrasts the positioning of the university as the ultimate global source of "universalized truths" (Frank and Meyer, *The University and the Global Knowledge Society*), on the one hand, and as an institution (and organizations) seriously challenged by neonationalism, populism, and reactionary forces that aim to curtail essential academic freedom, on the other (Douglass, *Neo-Nationalism and Universities*). Tensions between the global and national frames of reference are not new, as the university has been essential to the development of nation-state governance, to the diffusion of ideas by and influence of international organizations, and to the establishment and maintenance of global intellectual networks (Kennedy, *Globalizing Knowledge*; see also Kamola, *Making the World Global*, on the dispersion of knowledge production beyond the university, especially think tanks, the World Bank, and others). Kosmützky and Krücken ("Still the Century of the University as a Global Institution?") emphasize the importance of comparative perspectives on governance and on competition and collaboration on various levels in understanding contemporary patterns of university development and threats to collegiality and autonomy.

19. For example, some of Einstein's observations and theories have taken a century to prove, with decades of infrastructure development and billions of dollars invested in experiments such as the Laser Interferometer Gravitational-Wave Observatory (LIGO, "LIGO Lab"; see www.ligo.caltech.edu). This long-standing international research collaboration—a massive

research group of more than a thousand scientists working together worldwide—achieved their sought-after scientific discovery on September 14, 2015: detecting and confirming gravitational waves after half a century of comprehensive infrastructure development and tuning of intercontinental instruments to detect gravitational waves (Collins, *Gravity's Kiss*). See also the literature on microprocesses of discovery under the topic known as the sociology of scientific knowledge, e.g. Barnes, Bloor, and Henry, *Scientific Knowledge*.

20. Introductory sources in which science is considered a positive are Psillos, *Scientific Realism*; Ladyman, *Understanding Philosophy of Science*; Carnap, *An Introduction to the Philosophy of Science*.

21. Merton, *The Sociology of Science*; Merton and Barber, *The Travels and Adventures of Serendipity*; see also Ben-David and Sullivan, "Sociology of Science."

22. Joseph Ben-David analyzed the conditions of scientific growth historically and comparatively, emphasizing the ethos; the institutionalization; and the national traditions, policymaking, and conditions of science. In elaborating the multidimensional role of universities, he argued that already in the Middle Ages the European university had developed inexorable tensions related to the pedagogical transfer and ongoing production of new scientific knowledge: between the "formation of intellectual continuity and consensus, on the one hand, and dissent and revolution, on the other" (Ben-David, *Scientific Growth*, p. 305).

CHAPTER 3

1. About early scientific academies, see Stock, Powell, and Reisz, "Higher Education and Scientific Research in Germany"; Matthies and Stock, "Universitätsstudium und berufliches Handeln."

2. Selected histories of an evolving culture of science include Stichweh, *Wissenschaft, Universität, Professionen*, first and second editions; Novikoff, *The Medieval Culture of Disputation*.

3. See Crane, *Invisible Colleges*.

4. See Salsburg, *The Lady Tasting Tea*.

5. There are numerous rich historical studies on this topic, and Göttingen often figures prominently, such as those by Konrad Jarausch ("American Students in Germany, 1815–1914"), who reconstructs the students from Germany and the U.S. attending Göttingen University; William Clark's *Academic Charisma and the Origins of the Research University*, which provides an in-depth account of the sources of the modern research university, and its development especially in German cultural space, influenced by religions and the state; and James Axtell, *Wisdom's Workshop: The Rise of the Modern University*, Chapter 5. For recent examples of scholarship on the centrality of Germany (and Göttingen) in institutionalizing the research university, see Heinz-Dieter Meyer's *The Design of the University*; and Emily Levine's "Baltimore Teaches, Göttingen Learns" and *Allies and Rivals*. More broadly, and comparatively, see Rüegg, *Vom 19. Jahrhundert zum Zweiten Weltkrieg 1800–1945*; Ben-David, *Centers of Learning*; Lenhardt, *Hochschulen in Deutschland und den USA*.

6. Key texts on the subject have been collected and translated by Louis Menand, Paul Reitter, and Chad Wellmon, *The Rise of the Modern Research University*.

7. For an example of the break of several centuries in university institutionalization and contemporary developments in France, see Christine Musselin, *La Grande Course des Universités* and "Bringing Universities to the Centre of the French Higher Education System?"

8. The classic source in English on the reform of the medieval German university is Mc-Clelland, *State, Society, and University in Germany*.

9. Mitterle and Stock, "Higher Education Expansion in Germany"; Stock, Powell, and Reisz, "Higher Education and Scientific Research in Germany."

10. On milestones in the history of the Georgia Augusta University of Göttingen, see Georg-August-Universität Göttingen, "The History of the University," https://www.uni-goettingen.de/en/history+of+the+university/52652.html.

11. Miguel Urquiola (*Markets, Minds, and Money*, Table 1.4) notes that in Nobel Prize winners' biographies, Göttingen was among the top ten universities mentioned from 1855 through 1900 and fifth between 1901 and 1940, after only Cambridge, Harvard, Humboldt, and Columbia.

12. Useful historical analysis of this influential university can be found in Iggers, "The Intellectual Foundations of Nineteenth-Century 'Scientific' History"; Reid, *Hilbert-Courant*; Levine, "Baltimore Teaches, Göttingen Learns" and *Allies and Enemies*; Gongzhen, "Historical Study of the University of Göttingen"; Reisz, "Göttingen in Baltimore or the Americanization of the German University?"

13. For the origins and complicated history of the concept Realpolitik as it passed from mid-nineteenth-century Germany to contemporary Anglo-American usage, see Bew, *Realpolitik*.

14. Horlacher, *The Educated Subject and the German Concept of Bildung*.

15. See Ash, "Bachelor of What, Master of Whom?" And on the bidirectional influence of German and American research universities, see Levine, "Baltimore Teaches, Göttingen Learns" and *Allies and Rivals*; and Meyer, *The Design of the University*.

16. Edward Shils (*Max Weber on Universities*) collects Weber's writings, and powerfully prescient warnings, about universities, state power, and the dignity of the academic calling in Imperial Germany. These were forerunners to some of the ideas of Parsons, who translated one of Weber's books into English while he was himself at Heidelberg. Contemporarily, Frank and Gabler (*Reconstructing the University*) chart the vast shifts and differentiation in academic disciplines over the twentieth century, showing how the once-dominant humanities have declined in favor of the newer social sciences, with natural sciences relatively stable in the world's university curricula around the world.

17. vom Brocke, "Wege aus der Krise."; Stock, Powell, and Reisz, "Higher Education and Scientific Research in Germany."

18. Quote from Professor T. Maudlin of New York University in Musser, "A Defense of the Reality of Time."

19. The best source and compilation of statistics on historical enrollments and origins on the expansion of universities in German society is Paul Windolf's *Expansion and Structural Change*. Windolf's study is based on systematic statistical analyses of the most comprehensive set of enrollment data (more than that used in earlier histories). See also Fritz Ringer, *The Decline of the German Mandarins*, for a contrasting conclusion about expansion and integration of the university in Germany.

20. Ben-David, *Centers of Learning*.

21. Waldinger, "Peer Effects in Science."

22. See Bracher, *The German Dictatorship*.

CHAPTER 4

1. See, e.g., Levine, *Allies and Rivals*.

2. See Geiger's (*Research and Relevant Knowledge* and *The History of American Higher Education*) detailed histories of universities and research in the U.S. and his observation of the importance of a "knowledge conglomerate," an organizational implementation at universities of the American innovations on the university-science model.

3. For the empirical development of this argument, see Fernandez et al., "A Symbiosis of Access"; for related arguments, see Labaree, *A Perfect Mess*; and Brint, *Two Cheers for Higher Education*.

4. For foundational histories of the development of the American version of the research university and his concept of an emerging "knowledge conglomerate,"see Roger Geiger's *The History of American Higher Education* and his *Research and Relevant Knowledge*.

5. Williams (*Evan Pugh's Penn State*) provides a detailed history of Pugh and the early development of this land-grant university. See also Diehl, *Americans and German Scholarship, 1770–1870*.

6. Quote from a journalist's account of speeches by the educational reformer Harrison Howard circa 1840, on page 133 of Kett, *The Pursuit of Knowledge Under Difficulties*.

7. See Salsburg, *The Lady Tasting Tea*.

8. On this point, see Stevenson, "Scholarly Means to Evangelical Ends"; and Baker, "The Great Antagonism That Never Was." Plus, some large contemporary American research-intensive universities still explicitly cast their educational and knowledge production missions in this religio-cultural image; see, e.g., Baylor University's mission statement: https://illuminate.web.baylor.edu/about/baylors-academic-strategic-plan-illuminate.

9. A second *Morrill Land-Grant Act* in 1890 established public universities in the West and the South, as the first one did in the Northeast.

10. See a review of the lack of empirical evidence behind these common critiques in Baker, *The Schooled Society*.

11. For rich descriptions of some of the historical intricacies of American higher education in the nineteenth century, see Geiger, *The Land-Grant Act and American Higher Education*; and Williams, *Evan Pugh's Penn State*.

12. Geiger, *The History of American Higher Education*.

13. Rawlings and Bourgeois, "The Complexity of Institutional Niches."

14. For in-depth information on the development of science and technology in the U.S. and globally, including doctoral training, see National Science Board, "The State of U.S. Science and Engineering 2020."

15. For additional research on these points and trends described further on, see Mohrman, Ma, and Baker, "The Research University in Transition"; Baker, "Mass Higher Education and the Super Research University"; Baker and Lenhardt, "Privatization, Mass Higher Education, and the Super Research University"; and Baker, Fleishman, and Luo, "Symbiosis Among Access, Societal Support, and Scientific Productivity."

CHAPTER 5

1. This is not to ignore or discount the importance of key state-level plans for the development of large and differentiated higher education systems, most famously, Clark Kerr's sage and controversial 1960 California Master Plan for Higher Education. Among Kerr's most significant achievements at the helm of the University of California, the Master Plan reflected his vision of "the multiversity" in the U.S. in the second half of the twentieth century and his comprehensive understanding of the changing needs and future ambitions of

California (Kerr, *The Uses of the University*). Its implementation secured Californian higher education (inter)national prominence, and Kerr entered the pantheon of higher education thinkers and policymakers. He updated *The Uses of the University* in five editions to discuss new challenges to higher education, from the decline of liberal arts and dwindling resources to increasing students, use of information technology in teaching, and ethical concerns about biotechnology (on his legacy, see the contributions in Rothblatt, *Clark Kerr's World of Higher Education Reaches the 21st Century*).

2. Hunt, "The American Remission of the Boxer Indemnity."

3. See National Science Foundation, *Doctorate Recipients from U.S. Universities*.

4. Glass, Buus, and Braskamp, *Uneven Experiences*.

5. For informative technical analyses on these and related trends, see Ashkenas, Park, and Pearce, "Even with Affirmative Action, Blacks and Hispanics Are More Underrepresented at Top Colleges Than 35 Years Ago"; Fernandez et al., "A Symbiosis of Access." Also see Bowen and Rudenstine, *In Pursuit of the PhD*; and Posselt et al., "Access Without Equity."

CHAPTER 6

1. It also helped that Harnack was a personal friend of the kaiser's. For Harnack's intellectual impact, see Pauck, *Harnack and Troeltsch*, and for the early development of research and industry, see Rohrbeck, "F+E-Politik von Unternehmen." See Max-Planck-Gesellschaft, "Adolf Harnack's Memorandum for a Reform of German Science."

2. For a discussion of the German genius culture, see Watson, *The German Genius*; for its application to institutes that evidently made up the Max Planck Gesellschaft and the continuation of the Harnack Principle, see Peacock, "Academic Precarity as Hierarchical Dependence in the Max Planck Society"; for involvement in Nazi atrocities, see Heim, Sachse, and Walker, *The Kaiser Wilhelm Society Under National Socialism*; for post–World War II development, see Vierhaus, "Bemerkungen zum sogenannten Harnack-Prinzip Mythos und Realität.".

3. Bode, *Kommentierte Grafiken zum Deutschen Hochschul- und Forschungssystem*, p. 38, authors' translation; see also Larivière, et al., "Bibliometrics."

4. See Zippel, *Women in Global Science*, on gender bias in science.

5. Good sources on these points about the dual-pillar policy are Hinze, "Forschungsförderung und ihre Finanzierung"; Hohn, "Governance-Strukturen und institutioneller Wandel des außeruniversitären Forschungssystems Deutschlands"; Dusdal, *Welche Organisationsformen produzieren Wissenschaft?*; and Dusdal et al., "University vs. Research Institute?"

6. For description and analyses of the research university under the dual pillar policy, see Pritchard, "Trends in the Restructuring of German Universities"; Baker and Lenhardt, "The Institutional Crisis of the German Research University"; Henke and Pasternack, "Hochschulsystemfinanzierung"; Hüther and Krücken, *Higher Education in Germany*; Dusdal, *Welche Organisationsformen produzieren Wissenschaft?*; Kosmützky and Krücken, "Governing Research."

7. Musselin, *La Grande Course des Universités* and "Bringing Universities to the Centre of the French Higher Education System?"; Powell and Dusdal, "The European Center of Science Productivity"; Brankovic, Ringel, and Werron, "How Rankings Produce Competition."

8. In 1997, May ("The Scientific Wealth of Nations") challenged the benefit of this dual-pillar model in both Germany and France, despite their excellent researchers in research institutes, emphasizing that the "nonhierarchical nature of most North American and northern

European universities, coupled with the pervasive presence of irreverent young undergraduate and postgraduate students, could be the best environment for productive research."

9. Heinze and Kuhlmann, "Across Institutional Boundaries?"

10. See detailed analysis in Dusdal et al., "University vs. Research Institute?"

11. Urquiola, *Markets, Minds, and Money.*

12. Robin and Schubert, "Cooperation with Public Research Institutions and Success in Innovation"; Brint and Carr, "The Scientific Research Output of US Research Universities, 1980–2010"; Powell and Dusdal, "Science Production in Germany, France, Belgium, and Luxembourg."

13. Whereas funding levels across many countries have declined in recent decades, in Germany the research funds allocated to research institutes and large-scale targeted funding lines from national government ministries have increased opportunities in some fields; however, specific analyses of resource flows and personnel are rare (but see Whitley, Gläser, and Laudel, "The Impact of Changing Funding and Authority Relationships on Scientific Innovations"). Multilevel, longitudinal, and comparative research designs are needed to provide mechanism-based analysis that facilitates causal explanations adequate to explain contemporary multisource research funding and its impact on knowledge production; see Gläser and Serrano-Velarde, "Changing Funding Arrangements and the Production of Scientific Knowledge."

14. Because the obtained personnel counts of both pillars included scientists at all levels of career development, from postdoctoral fellows and PhD technicians to senior investigators with established labs, there is likely a wide dispersion around these mean ratios. Plus, these estimates in many cases include technical personnel not routinely included as coauthors of papers; worldwide, across all STEM+ domains, most regularly authoring scientists, including those in Germany, publish an average of two to two-and-a-half papers per year. For related analyses, see Helmich, Gruber, and Frietsch, *Performance and Structures of the German Science System 2017*; Wang and Barabási, *The Science of Science.*

15. East Germany, the former German Democratic Republic (GDR), adopted a state-controlled system of research institutes in an Academy of Sciences in line with USSR science policy and did not contribute much to Western STEM+ journals. This provided suboptimal conditions for scientific productivity (see Gläser and Meske, *Anwendungsorientierung von Grundlagenforschung?*; Mayntz, "Academy of Sciences in Crisis"). Between 1950 and 1990, the total growth rate in the STEM+ paper publication of East and West Germany was due mostly to West Germany–based universities and institutes. After the Cold War, the GDR's Academy of Sciences was transformed and the Western German dual-pillar model implemented. As Lovakov, Chankseliani, and Panova ("Universities vs. Research Institutes?") show in their analysis of formerly Soviet countries (Armenia, Azerbaijan, Belarus, Estonia, Georgia, Kazakhstan, Kyrgyzstan, Latvia, Lithuania, Moldova, Russia, Tajikistan, Ukraine, Uzbekistan), the number of research institutes dwarfs the number of universities, originally considered to be mainly teaching organizations, but in the decades since universities' research output has grown higher than that of institutes, even if in most of these national contexts the Academies of Sciences remain important centers of research.

16. More research is needed on the importance of disciplinary networks and patterns of competition between and collaboration across organizational forms (e.g., Dusdal, *Welche Organisationsformen produzieren Wissenschaft?*; Dusdal, Oberg, and Powell, "Das Verhältnis zwischen Hochschule und Wissenschaft in Deutschland"; Dusdal and Powell, "Benefits, Motivations, and Challenges of International Collaborative Research."

17. Zapp and Powell, "Moving Towards Mode 2?"; Marques, Zapp, and Powell, "Europeanizing Universities"; Kosmützky and Krücken, "Governing Research."

18. Besio, *Forschungsprojekte*; Torka, *Die Projektförmigkeit der Forschung.*

19. Münch, *Die akademische Elite*; Leibfried, *Die Exzellenzinitiative.*

20. Buenstorf and Koenig, "Interrelated Funding Streams in a Multi-Funder University System."

21. Musselin, "New Forms of Competition in Higher Education"; Powell, "Higher Education and the Exponential Rise of Science"; Krücken, "Multiple Competitions in Higher Education."

22. Hottenrott and Lawson, "A First Look at Multiple Institutional Affiliations."

23. See, e.g., Godin and Gingras, "The Place of Universities in the System of Knowledge Production"; Schofer and Meyer, "The Worldwide Expansion of Higher Education in the Twentieth Century"; Frank and Meyer, *The University and the Global Knowledge Economy.*

CHAPTER 7

1. See, e.g., Leydesdorff and Wagner, "Is the United States Losing Ground in Science?"; Xie and Killewald, *Is American Science in Decline?*; Viglione, "China Is Closing the Gap with the United States on Research Spending."

2. Marginson and Xu, *Changing Higher Education in East Asia.*

3. "We were crazy, crazy about work, I was blinded. All I could see was whether I could make Korea stand in the center of the world through this research." South Korean scientist Hwang Woo-Suk's statement about his fraudulent findings (McCurry, "Disgrace"). And on motivational practices, see Abritis, McCook, and Retraction Watch, "Cash Incentives for Papers Go Global."

4. Lee, "Education Hubs and Talent Development."

5. For detailed accounts of educational and scientific development in each of these countries by collaborators in the SPHERE Project, see chapters by Zhang, Sun, and Bao, "The Rise of Higher Education and Science in China;" Shima, "Changing Science Production in Japan;" Fu, "Science Production in Taiwanese Universities, 1980–2011;" and Kim and Choi, "The Growth of Higher Education and Science Production in South Korea Since 1945," all from Powell, Baker, and Fernandez, *The Century of Science.*

6. For statistical trends in education for these countries, see OECD Indicators, "Education at a Glance 2022, with a Spotlight on Tertiary Education" (2022 and previous years): https://www.oecd.org/education/education-at-a-glance/.

7. Oleksiyenko, "On the Shoulders of Giants?"; Zhang, Sun, and Bao, "The Rise of Higher Education and Science in China."

8. For more on this idea and the development of a global model of the research university, see Merton, "The Matthew Effect in Science, II"; and Mohrman, Ma, and Baker, "The Research University in Transition." For evaluations in Taiwan and China, respectively, see Fu, "Science Production in Taiwanese Universities, 1980–2011"; and Zhang, Patton, and Kenney, "Building Global-Class Universities." And for Japan and Korea, see Shima, "The Changing Science Production in Japan"; Kim and Choi, "The Growth of Higher Education and Science Production in South Korea Since 1945"; and Nam, "An Assessment of the Impact of the Center of Excellence Program on the Research Production of Korean Universities from 1989 to 2011."

9. On "peak massification" and the coming demographic challenges and financial con-

straints in the regions' expanded higher education systems, see Horta, "Emerging and Near Future Challenges of Higher Education in East Asia."

10. See Fu, Baker, and Zhang, "Engineering a World Class University?"

11. Shima, "Changing Science Production in Japan."

12. For extensive analysis of this history, see Yu, "The Collaboration Network of American and Korean Universities." And for a more cultural analysis of education in South Korea, see Schaub et al., "Policy Reformer's Dream or Nightmare?"

CHAPTER 8

1. See Beydoun and Baum, *The Glass Palace*; on Qatar's breakneck development, see Fromherz, *Qatar*.

2. See Crist, "Innovation in a Small State" and "'A Fever of Research'."

3. Founded in 1995, The Qatar Foundation proclaims, "Our Manifesto: We gathered a generation of big ideas, Planted the roots of knowledge, Nurtured the power of a thought. We prepared for the future, and, in time, found ourselves living it. In our world, Big ideas flow, Knowledge grows, and Thoughts roam readily. Delighted by the luminance of innovation, We spark the kindles of a generation's finest wonders. We debate. We discover. We create. We curate. We think. We transform. And we do what we always did best, We unlock human potential" (see Qatar Foundation, "About Qatar Foundation").

4. Halabi et al., "Preferential Allele Expression Analysis Identifies Shared Germline and Somatic Driver Genes in Advanced Ovarian Cancer."

5. STATEC, "626,000 Inhabitants as of January 1, 2020."

6. Graf and Gardin, "Transnational Skills Development in Post-Industrial Knowledge Economies."

7. See International Monetary Fund (IMF) database: https://data.imf.org/.

8. Braband and Powell, "European Embeddedness and the Founding of Luxembourg's 21st Century Research University."

9. Rohstock and Schreiber, "The Grand Duchy on the Grand Tour."

10. Two decades after its founding, Qatar Foundation's Education City campus in Doha hosts a dozen branch campuses of well-known universities, including Carnegie Mellon University Qatar, Weill Cornell Medicine—Qatar, Georgetown University School of Foreign Service in Qatar, Northwestern University in Qatar, Texas A&M University in Qatar, Virginia Commonwealth University Qatar, and HEC Paris in Qatar; University College London Qatar closed in 2020. Qatar Foundation's homegrown Hamad bin Khalifa University is an emerging research-oriented university offering postgraduate multidisciplinary courses that respond to the region's specific needs, housing institutes and centers, such as the Qatar Faculty of Islamic Studies. Qatar Foundation aims to place Qatar among leading countries in scientific and biomedical research. Facilitating this, the Qatar National Research Fund supports Qatar's most talented scientists and their collaborators around the world with the resources they need. Further infrastructure includes Qatar Science & Technology Park and Sidra Medical Center. In 2009, Sheikha Moza established The World Innovation Summit for Education (WISE) an international conference to scope the future of education.

11. Hennicot-Schoepges, "Translation—Erna Hennicot-Schoepges."

12. Hennicot-Schoepges, "Génèse d'un défi."

13. European Commission, "Communication from the Commission to the Council, the European Parliament, the Economic and Social Committee and the Committee of the Re-

gions." https://eur-lex.europa.eu/LexUriServ/LexUriServ.do?uri=CELEX:52000DC0379:EN:HTML. The emerging European model in skill formation was a bricolage of the most influential national models, especially in higher education: German, British, French, and American, see Powell, Bernhard, and Graf, "The Emergent European Model in Skill Formation."

14. In the Ministry, Germain Dondelinger, who had been educated in England, oversaw the massive Belval project and also facilitated the establishment of the (short-lived) Max Planck Institute, specializing in European and international law, whose resources were folded into the University of Luxembourg in 2024. On early developments of Luxembourg's R&D infrastructure and scientific culture, see Meyer, "The Dynamics of Science in a Small Country" and "Creativity and Its Contexts." On the reform of higher education governance in Luxembourg, see Harmsen and Powell, "Higher Education Systems and Institutions, Luxembourg."

15. See Rohstock, "Wider die Gleichmacherei! Luxemburgs langer Weg zur Universität 1848–2003"; Hennicot-Schoepges, "Génèse d'un défi"; Braband and Powell, "European Embeddedness and the Founding of Luxembourg's 21st Century Research University."

16. Frank and Meyer, *The University and the Global Knowledge Economy*, p. 142.

17. Jack Lee ("Education Hubs and Talent Development") emphasizes that while Malaysia, Singapore, and Hong Kong are education hubs that share three distinct objectives—of developing local talent, attracting foreign talent, and repatriating diasporic talent—the relative emphasis they give to these objectives differs. Luxembourg and Qatar policymakers have emphasized the first two objectives. As Ortiga, Chou, and Wang ("Competing for Academic Labor") emphasize, even a hub as strong as Singapore, with multiple formidable research universities, must overcome numerous challenges to successfully recruit academic labor from elsewhere.

18. Stevens, Miller-Idriss, and Shami, *Seeing the World*.

19. Klemenčič, "Epilogue: Reflections on a New Flagship University"; Kamola (*Making the World Global*) uncovers American universities' production of knowledge about the world in terms of nation-states and market-oriented globalization as the main logic driving change in U.S. higher education. In Luxembourg, the state provides postsecondary education as a public good, charging negligible tuition fees and providing grants for students wishing to study abroad (a requirement of all BA students at the University of Luxembourg), thus continuing the tradition of outgoing mobility as decreed by the parliament (Kmiotek-Meier, Karl, and Powell, "Designing the (Most) Mobile University"; Kmiotek-Meier and Powell, "Evaluating Universal Student Mobility."

20. Powell, "International National Universities."

21. The UGR consists of universities in four countries, with over ten thousand researchers and over seven thousand doctoral candidates who study in French, German, and English at the Universities of Kaiserslautern-Landau, Liège, Lorraine, Luxembourg, Saarland, and Trier, all located within the "Greater Region" of Germany (Rhineland-Palatinate and Saarland), Belgium (Wallonia), France (Grand-Est Region–Lorraine), and the Grand Duchy of Luxembourg in the center; see www.uni-gr.eu.

22. Schofer and Meyer, "The Worldwide Expansion of Higher Education in the Twentieth Century."

23. Estimate for the world in 2018 from UIS, "Welcome to UIS.Stat," http://data.uis.unesco.org. See also Frank and Meyer, *The University and the Global Knowledge Economy*; Schofer, Ramirez, and Meyer, "The Societal Consequences of Higher Education."

24. Kirby and van der Wende, "The New Silk Road"; Marginson, "'All Things Are in Flux.'"

25. About universities and science in Brazil, see Cross, Thomson, and Sinclair, "Research in Brazil."

CHAPTER 9

1. Built from 2004 through 2010 with funds from various sources and coordinated by the University of Wisconsin–Madison, the IceCube South Pole Neutrino Observatory is the first "detector" of its kind, enabling scientists to observe the cosmos from deep within the ice at the South Pole. Using this unique observatory, the IceCube Collaboration is a worldwide group of scientists searching for and measuring neutrinos, nearly massless subatomic particles (see IceCube Neutrino Observatory, "IceCube," https://icecube.wisc.edu/).

2. The IceCube is based at the Amundsen-Scott South Pole Station, located on the high plateau of Antarctica and administered by the U.S. National Science Foundation's United States Antarctic Program (USAP): https://www.usap.gov/. It is named to honor Norwegian Roald Amundsen and Briton Robert F. Scott, who led separate teams to reach the pole: Amundsen's team arrived first, on December 14, 1911; Scott's team arrived second, on January 17, 1912.

3. In early 2021, the IceCube project was awarded the Bruno Rossi Prize by the High Energy Astrophysics Division of the American Astronomical Society, the largest professional organization of astronomers in the United States.

4. Astrophysical models of particle-accelerating, black-hole-powered blazars predicted that they ought to emit both light and neutrinos, and the empirical evidence from IceCube confirms this. In November 2022, using data recorded with the IceCube neutrino detector between 2011 and 2020, neutrino emissions from astrophysical objects were reported from NGC 1068, a galaxy just forty-seven million light-years away (IceCube Collaboration, "Evidence for Neutrino Emission from the Nearby Active Galaxy NGC 1068").

5. Jones, Wuchty and Uzzi, "Multi-University Research Teams."

6. Running from 1990 through 2003, The Human Genome Project was a vast international scientific undertaking that identified and cataloged all base pairs that make up human DNA and mapped the approximately 20,500 genes of the human genome (see National Human Genome Research Institute, "What Is the Human Genome Project?": https://www.genome.gov/human-genome-project/What).

7. Bozeman and Boardman, "Assessing Research Collaboration Studies"; see also Beaver, "Reflections on Scientific Collaboration (and Its Study)"; Geiger and Sá, *Tapping the Riches of Science.*

8. For over half a century, the field of bibliometrics has developed to investigate such patterns and trends of scientific collaboration and copublication across boundaries (see Powell, "Learning from Collaboration"; Glänzel and Schubert, "Analysing Scientific Networks Through Co-Authorship," Glänzel, "National Characteristics in International Scientific Co-Authorship Relations"; Glänzel, "Seven Myths in Bibliometrics About Facts and Fiction in Quantitative Science Studies"); Sugimoto and Larivière, *Measuring Research.*

9. Kosmützky, "A Two-Sided Medal"; Wöhlert, "Communication in International Collaborative Research Teams"; Kosmützky and Wöhlert, "Varieties of Collaboration"; Dusdal and Powell, "Benefits, Motivations, and Challenges of International Collaborative Research."

10. On the limitations of coauthored papers as a measure of collaboration, see Laudel, "What Do We Measure by Co-Authorships?"

11. See Hicks and Katz, "Where Is Science Going?"; Adams, "The Fourth Age of Research"; Mosbah-Natanson and Gingras, "The Globalization of Social Sciences?"; Powell

et al., "Introduction: The Worldwide Triumph of the Research University and Globalizing Science."

12. See IceCube Collaboration: Abbasi, et al. "Detection of Astrophysical Tau Neutrino Candidates in IceCube."

13. At perhaps the most famous international collaborative infrastructure, CERN, the European Organization for Nuclear Research a hundred meters underground below the Franco–Swiss border, operates the world's largest particle physics laboratory, for researchers from twenty-three member states, the huge increase in number of authors occurred even earlier, from dozens of authors in the 1980s to hundreds (Gillies, *CERN and the Higgs-Boson*, p. 114).

14. Adams, "Collaborations: The Rise of Research Networks" and Adams, Pendlebury, Potter, and Szomszor, "Global Research Report," on the basis of Clarivate Analytics's Web of Science, identified the consequences of collaboration and the rising numbers of research articles with a thousand or more unique authors across more than one hundred different countries. Such complex combinations of authors, organizational affiliations, and host countries gives rise to patterns unlike typical academic papers. The report notes general increases of "multi-authorship" (more than ten authors; more than five countries) and as well as "hyper-authorship" (more than a hundred authors; more than thirty countries). In the complete Web of Science, the median number of coauthors is three. Ninety-five percent of global output has ten or fewer authors; the median number of countries on an article is one, and 99 percent of global output has authors from five or fewer countries. Complex authorship (many authors, many countries) has continued to rise over the past years.

15. Aksnes and Sivertson, "Global Trends in International Research Collaboration, 1980–2021."

16. As Vincent Larivière and colleagues show in their longitudinal analysis (1900–2011), teams that are larger and more diverse (researchers from more organizations and more countries participating) are necessary to realize higher impact, as measured by citations. At the same time, "constant inflation of collaboration since 1900 has resulted in diminishing citation returns" (see Larivière et al., "Team Size Matters"). There is also new science on characteristics of teams and the kinds of scientific advances they create; see, e.g., Wu, Wang, and Evans, "Large Teams Develop and Small Teams Disrupt Science and Technology."

17. See Dusdal, *Welche Organisationsformen produzieren Wissenschaft?*; Dusdal, Oberg, and Powell, "Das Verhältnis zwischen Hochschule und Wissenschaft in Deutschland," on disciplinary and organizational form differences.

18. See publication list at https://icecube.wisc.edu/science/publications.

19. On the influence of the European Union policies on the construction and expansion of university networks and their impact, see Zapp, Marques, and Powell, *European Educational Research (Re)Constructed*; Marques, Zapp, and Powell, "Europeanizing Universities"; Marques and Graf, "Pushing Boundaries."

20. Yu, "The Collaboration Network of American and Korean Universities."

21. For further in-depth national comparisons, see Aksnes and Sivertson, "Global Trends in International Research Collaboration, 1980–2021."

22. National research portfolios are evolving to become more diverse, even as global science as a whole has become more specialized over the past decades, with differences between disciplines and clusters of natural, physical, and social scientific disciplines; see Miao et al., "The Latent Structure of Global Scientific Development."

CHAPTER 10

1. Stigler, "The Intellectual and the Marketplace."

2. See Baker, *The Schooled Society,* for a review of the historical and sociological research on this general interpretation of the university in society.

3. See Labaree, *A Perfect Mess.*

4. The approximately three hundred are those noted in Chapter 4, all in the three highest categories of the Carnegie Classification, and four hundred additional postsecondary institutions were identified in our SPHERE project publication database as producing at least ten STEM+ papers annually in indexed journals.

5. Baker and Lenhardt, "The Institutional Crisis of the German Research University"; Powell and Solga, "Why Are Higher Education Participation Rates in Germany So Low?"; Mitterle and Stock, "Higher Education Expansion in Germany."

6. Alternatives to peer review, such as allocation of research funds by lottery, are controversial and hardly tested; see Barlösius and Philipps, "Random Grant Allocation from the Researchers' Perspective."

7. See Urquiola, *Markets, Minds, and Money.*

8. On the institutional and organizational conditions for scientific breakthroughs, see, e.g., Hollingsworth, "The Role of Institutions and Organizations in Shaping Radical Scientific Innovations"; Hollingsworth and Hollingsworth, *Major Discoveries, Creativity, and the Dynamics of Science;* Heinze, von der Heyden, and Pithan, "Institutional Environments and Breakthroughs in Science."

9. See, e.g., Drori et al., *Science in the Modern World Polity.*

10. As explained in the data note in the Preface, for all the analyses and particularly those prior to 1980, these indicators represent main scientific journals and the faculty-scientists at universities that published them. And while they provide an accurate view of the trends, because they are based on sampled and weighted estimates they should not be considered a census or full accounting of the absolute total since some unknown number of STEM+ publications, authors, and institutions are unaccounted for at each point.

11. See The World Bank, "School Enrollment, Tertiary (% Gross)": https://data.worldbank.org/indicator/SE.TER.ENRR

12. See, e.g., Cowen, *The Great Stagnation;* Horgan, *The End of Science;* Ness, *The Creativity Crisis.*

13. Marques, "Research Governance Through Public Funding Instruments"; Marques, "Governing European Educational Research Through Ideas?"

14. Flink, "Taking the Pulse of Science Diplomacy and Developing Practices of Valuation"; Epping, *Exploring the Institutionalisation of Science Diplomacy.*

15. Dusdal and Powell, "Benefits, Motivations, and Challenges of International Collaborative Research."

16. The Royal Society, *Knowledge, Networks, and Nations;* Sugimoto et al., "Scientists Have Most Impact When They're Free to Move"; Powell, "Higher Education and the Exponential Rise of Science"; Wagner, *The Collaborative Era in Science.*

17. On the LIGO collaborative, see Collins, *Gravity's Kiss.*

18. In the COVID-19 pandemic, interorganizational form collaboration has again been essential in the biotech industry (see Powell, Koput, and Smith-Doerr, "Interorganizational Collaboration and the Locus of Innovation"), even if such partnerships are not without their challenges given different goals and modes of operation (see Evans, "Industry Induces Academic Science to Know Less About More"). Jason Owen-Smith (*Research Universities and*

the Public Good, Chapter 3) argues that a strength of the U.S.'s complex and decentralized research system is the diversity of the university collaboration networks that emerge, but public investment is necessary throughout to sustain it.

19. Kuhn, *The Structure of Scientific Revolution*.

20. Austrialia's scientists' extensive publishing and their Anglophone advantage—English language fluency and considerable exchange and collaboration among English-speaking countries—here puts them in the first group.

21. Using Scopus data for science and engineering disciplines, the U.S. National Center for Science and Engineering Statistics reports that in 2020, the top four producing countries were China 669,744 S&E papers (22.7 percent of world total); U.S. 455,856 (15.5 percent); India 149,213 (5 percent); and Germany 109,379 (3.7 percent).

22. See Baker, *The Schooled Society*, for review of education effects on the economy.

23. Zapp, Marques, and Powell, *European Educational Research (Re)Constructed*.

24. This was manifested in the forced move of the Central European University from Budapest to Vienna, but is a much broader phenomenon that has also reached the U.S. On the weakening of the liberal global order via nationalism and populism and its impact on higher education worldwide through limits on academic freedom and even terrorist attacks, see Schofer, Lerch, and Meyer, "Illiberal Reactions to Higher Education." Support for science similarly suffers from populism; see Zapp, "The Legitimacy of Science and the Populist Backlash."

25. Zhang, "Understanding Chinese Science."

26. Powell et al., "Introduction: The Worldwide Triumph of the Research University and Globalizing Science."

Bibliography

Abritis, Alison, Alison McCook, and Retraction Watch. "Cash Incentives for Papers Go Global." *Science* 357, no. 6351 (August 11, 2017): 541. https://doi.org/10.1126/science.357.6351.541.

Abrutyn, Seth. "Toward a General Theory of Institutional Autonomy." *Sociological Theory* 27, no. 4 (2009): 449–65.

———. *Revisiting Institutionalism in Sociology: Putting the "Institution" Back in Institutional Analysis.* New York: Routledge, 2013.

Abrutyn, Seth, and Jonathan H. Turner. "The Old Institutionalism Meets the New Institutionalism." *Sociological Perspectives* 54, no. 3 (2011): 283–306. https://doi.org/10.1525/sop.2011.54.3.283.

Acemoglu, Daron, and David Autor. "Chapter 12—Skills, Tasks, and Technologies: Implications for Employment and Earnings." In *Handbook of Labor Economics,* ed. David Card and Orley Ashenfelter, 1043–1171. Vol. 4, Part B. Amsterdam: Elsevier, 2011. https://doi.org/10.1016/S0169-7218(11)02410-5.

———. "What Does Human Capital Do? A Review of Goldin and Katz's *The Race Between Education and Technology.*" *Journal of Economic Literature* 50, no. 2 (2012): 426–63. http://www.aeaweb.org/articles.php?doi=10.1257/jel.50.2.426.

Adams, James D. "Is the U.S. Losing Its Preeminence in Higher Education?" *NBER Working Paper Series* No. 15233, National Bureau of Economic Research, Cambridge, MA, 2009. https://doi.org/10.3386/w15233.

Adams, Jonathan. "The Rise of Research Networks." *Nature* 490 (2012): 335–36. doi: 10.1038/490335a.

———. "The Fourth Age of Research." *Nature* 497, (2013): 557–60. doi: 10.1038/497557a.

———. "Global Research Report: United Kingdom." Philadelphia: Thomson Reuters, 2011. https://www.researchgate.net/publication/276594967_Worldwide_open_access_UK_leadership/fulltext/55d8140508ae9d65948db25d/Worldwide-open-access-UK-leadership.pdf (accessed May 25, 2022).

Adams, Jonathan, David Pendlebury, Ross Potter, and Martin Szomszor. "Global Research Report: Multi-Authorship and Research Analytics." Philadephia: Institute for Scientific Information (ISI), 2019. https://investigacion.utem.cl/wp-content/uploads/documentos/documento150/ISI%20Global%20Research%20Report%20Multiautorship%20and%20research%20analytics.pdf (accessed May 30, 2022).

Aksnes, Dag W., and Gunnar Sivertsen. "Global Trends in International Research Collaboration, 1980–2021."*Journal of Data and Information Science* 8, no. 2 (2023): 26–42. https://doi.org/10.2478/jdis-2023-0015.

Allik, Jüri, Kalmer Lauk, and Anu Realo. "Factors Predicting the Scientific Wealth of Nations." *Cross-Cultural Research* 54, no. 4 (2020): 364–97. doi: 10.1177/1069397120910982.

Ash, Mitchell G. "Bachelor of What, Master of Whom? The Humboldt Myth and Historical

Transformations of Higher Education in German-Speaking Europe and the US." *European Journal of Education* 41, no. 2 (2006): 245–67. https://doi.org/10.1111/j.1465-3435.2006.00258.x.

Ashkenas, Jeremy, Haeyoun Park, and Adam Pearce. "Even with Affirmative Action, Blacks and Hispanics Are More Underrepresented at Top Colleges than 35 Years Ago." *The New York Times*, August 24, 2017.

Axtell, James. *Wisdom's Workshop: The Rise of the Modern University*. Princeton: Princeton University Press, 2016.

Baker, David P. "The Great Antagonism That Never Was: Unexpected Affinities Between Religion and Education in Post-Secular Society." *Theory and Society* 48, no. 1 (2019): 39–65. https://doi.org/10.1007/s11186-018-09338-w.

———. "Mass Higher Education and the Super Research University." *International Higher Education* 49 (2008): 9–10.

———. *The Schooled Society: The Educational Transformation of Global Culture*. Stanford, CA: Stanford University Press, 2014.

———. "Theoretische Zugänge zur Bildungsrevolution in der post-industriellen Gesellschaft: Mythen, Fakten und die Akademisierung der Berufsrollen." In *Akademisierung—Professionalisierung. Zum Verhältnis von Hochschulbildung, akademischem Wissen und Arbeitswelt*, ed. Alexander Mitterle, Annemarie Matthies, Annett Maiwald, and Christoph Schubert. Wiesbaden: Springer, in press.

Baker, David P., Shannon Smythe Fleishman, and Yuan Luo. "Symbiosis Among Access, Societal Support, and Scientific Productivity: Land-Grant Universities in the Late 20th Century." Paper presented at Penn State University conference on The Legacy and Promise: 150 Years of Land Grant Universities, University Park, PA, June 2011.

Baker, David P., and Gero Lenhardt. "The Institutional Crisis of the German Research University." *Higher Education Policy* 21 (2008): 49–64. https://doi.org/10.1057/palgrave.hep.8300178.

———. "Privatization, Mass Higher Education, and the Super Research University. Symbiotic or Zero-Sum Trends?" *Die Hochschule* 17, no. 2 (2008): 36–52.

Baker, David P., Maryellen Schaub, Junghee Choi, and Karly Ford. "Education: The Great Equalizer, Social Reproducer, or Legitimator of New Forms of Social Stratification?" In *The International Handbook of Sociology of Education*, ed. Mark Berends, Barbara Schneider, and Stephen Lamb. London: SAGE, in press.

Ball, Robert, and Dirk Tunger. *Bibliometrische Analysen—Daten, Fakten und Methoden. Grundwissen Bibliometrie für Wissenschaftler, Wissenschaftsmanager, Forschungseinrichtungen und Hochschulen. Schriften des Forschungszentrums Jülich*, vol. 12. Jülich, Germany: Forschungszentrum Jülich, 2005.

Barlösius, Eva, and Alex Philipps. "Random Grant Allocation from the Researchers' Perspective." *Social Science Information* 61, no. 1. (2022): 154–78. doi: 10.1177/05390184221076627.

Barnes, Barry, David Bloor, and John Henry. *Scientific Knowledge: A Sociological Analysis*. Chicago: University of Chicago Press, 1996.

Beaver, Donald D. "Reflections on Scientific Collaboration (and Its Study)." *Scientometrics* 52, no. 3 (2001): 365–77. https://doi.org/10.1023/A:1014254214337.

Ben-David, Joseph. *Centers of Learning: Britain, France, Germany, United States*. New Brunswick, NJ: Transaction, [1977] 2009.

———. *Scientific Growth: Essays on the Social Organization and Ethos of Science*, ed. Gad Freudenthal. Berkeley: University of California Press, 1991.

Ben-David, Joseph, and Teresa Sullivan. "Sociology of Science." *Annual Review of Sociology* 1 (1975): 203–22. https://doi.org/10.1146/annurev.so.01.080175.001223.

Besio, Christina. *Forschungsprojekte. Zum Organisationswandel in der Wissenschaft.* Bielefeld, Germany: Transcript, 2009.

Bew, John. *Realpolitik.* Oxford, UK: Oxford University Press, 2016.

Beydoun, Nasser M., and Jennifer Baum. *The Glass Palace: Illusions of Freedom and Democracy in Qatar.* New York: Algora, 2012.

Bode, Christian. *Kommentierte Grafiken zum Deutschen Hochschul- und Forschungssystem.* Bonn: DAAD, 2015.

Bornmann, Lutz, and Rüdiger Mutz. "Growth Rates of Modern Science: A Bibliometric Analysis Based on the Number of Publications and Cited References." *Journal of the Association for Information Science and Technology* 66, no. 11 (2015): 2215–22. https://doi.org/10.1002/asi.23329.

Bornmann, Lutz, Caroline Wagner, and Loet Leydesdorff. "BRICS Countries and Scientific Excellence: A Bibliometric Analysis of Most Frequently Cited Papers." *Journal of the Association for Information Science and Technology* 66, no. 7 (2015): 1507–13. https://doi.org/10.1002/asi.23333.

Bowen, William G., and Neil L. Rudenstine. *In Pursuit of the PhD.* Princeton, NJ: Princeton University Press, 1992.

Bozeman, Barry, and Craig Boardman. "Assessing Research Collaboration Studies: A Framework for Analysis." In *Research Collaboration and Team Science,* Barry Bozeman and Craig Boardman, 1–11. Cham, Switzerland: Springer, 2014.

Braband, Gangolf, and Justin J.W. Powell. "European Embeddedness and the Founding of Luxembourg's 21st Century Research University." *European Journal of Higher Education* 11, no. 3 (2021): 255–72. https://doi.org/10.1080/21568235.2021.1944251.

Bracher, Karl Dietrich. *The German Dictatorship: The Origins, Structure, and Consequences of National Socialism.* Harmondsworth, UK: Penguin, 1973.

Brainard, Jeffrey. "Scientists Are Drowning in COVID-19 Papers. Can New Tools Keep Them Afloat?" *Science* 13, no. 10 (2020): 1126. https://doi.org/10.1126/science.abc7839.

Brankovic, Jelena, Leopold Ringel, and Tobias Werron. "How Rankings Produce Competition: The Case of Global University Rankings." *Zeitschrift für Soziologie* 47 (2018): 270–87. https://doi.org/10.1515/zfsoz-2018-0118.

Brint, Steven. "An Institutional Geography of Knowledge Exchange: Producers, Exports, Imports, Trade Routes, and Metacognitive Metropoles." In *Education in a New Society: Renewing the Sociology of Education,* ed. Jal Mehta and Scott Davies, 115–43. Chicago: University of Chicago Press, 2018.

———. *Two Cheers for Higher Education: Why American Universities Are Stronger Than Ever—And How to Meet the Challenges They Face.* Princeton, NJ: Princeton University Press, 2019.

Brint, Steven, and Cynthia E. Carr. "The Scientific Research Output of US Research Universities, 1980–2010: Continuing Dispersion, Increasing Concentration, or Stable Inequality?" *Minerva* 55 (2017): 435–57. https://doi.org/10.1007/s11024-017-9330-4.

Buenstorf, Guido, and Johannes Koenig. "Interrelated Funding Streams in a Multi-Funder University System: Evidence from the German Exzellenzinitiative." *Research Policy* 49, no. 3 (2020): 103924. https://doi.org/10.1016/j.respol.2020.103924.

Carnap, Rudolf. *An Introduction to the Philosophy of Science.* North Chelmsford, MA: Courier Corporation, 2012.

Clark, William. *Academic Charisma and the Origins of the Research University*. Chicago: University of Chicago Press, 2006.

Collins, Harry. *Gravity's Kiss: The Detection of Gravitational Waves*. Cambridge, MA: MIT Press, 2017.

Cowen, Tyler. *The Great Stagnation: How America Ate All the Low-Hanging Fruit of Modern History, Got Sick, and Will (Eventually) Feel Better*. New York: Dutton, 2011.

Crane, Diana. *Invisible Colleges: Diffusion of Knowledge in Scientific Communities*. Chicago: University of Chicago Press, 1972. https://doi.org/10.1111/muwo.12082.

Crist, John T. "'A Fever of Research': Scientific Journal Article Production and the Emergence of a National Research System in Qatar, 1980–2011." In *The Century of Science: The Global Triumph of the Research University*, ed. Justin J.W. Powell, David P. Baker, and Frank Fernandez, 227–48. Bingley, UK: Emerald, 2017. https://doi.org/10.1108/S1479-367920170000033011.

———. "Innovation in a Small State: Qatar and the IBC Cluster Model of Higher Education." *The Muslim World* 105 (2015): 93–115.

Cross, Di, Simon Thomson, and Alexandra Sinclair. "Research in Brazil: A Report for CAPES by Clarivate Analytics." Clarivate Analytics, 2018. http://www.sibi.usp.br/wp-content/uploads/2018/01/Relat%C3%B3rio-Clarivate-Capes-InCites-Brasil-2018.pdf (accessed May 25, 2022).

Czaika, Mathias, and Sultan Orazbayev. "The Globalisation of Scientific Mobility, 1970–2014." *Applied Geography* 96 (2018): 1–10. https://doi.org/10.1016/j.apgeog.2018.04.017.

Diamond, Jared. *Guns, Germs, and Steel: The Fates of Human Societies*. New York: W.W. Norton, 1997.

Diehl, Carl. *Americans and German Scholarship, 1770–1870*. New Haven: Yale University Press, 1978.

Dong, Yuxiao, Hao Ma, Zhihong Shen, and Kuansan Wang. "A Century of Science: Globalization of Scientific Collaborations, Citations, and Innovations." *Proceedings of KDD '17*, August 13–17, 2017, Halifax, NS, Canada. doi: 10.1145/3097983.3098016.

Douglass, John Aubrey, ed. *Neo-Nationalism and Universities: Populists, Autocrats, and the Future of Higher Education*. Baltimore: Johns Hopkins University Press, 2021. doi:10.1353/book.85165.

Drori, Gili S., and John W. Meyer. "Global Scientization: An Environment for Expanded Organization." In *World Society and Organizational Change*, ed. Gili S. Drori, John W. Meyer, and Hokyu Hwang, 50–68. Oxford, UK: Oxford University Press, 2006.

———. Scientization: Making a World Safe for Organization. In *Transnational Governance. Institutional Dynamics of Regulation*, ed. Marie-Laure Djelic and Kerstin Sahlin-Andersson, 31–52. Oxford, UK: Oxford University Press, 2006.

Drori, Gili S., John W. Meyer, Francisco O. Ramirez, and Evan Schofer. *Science in the Modern World Polity: Institutionalization and Globalization*. Stanford, CA: Stanford University Press, 2003.

Dusdal, Jennifer. *Welche Organisationsformen produzieren Wissenschaft? Zum Verhältnis von Hochschule und Wissenschaft in Deutschland*. Frankfurt am Main, Germany: Campus Verlag, 2018.

Dusdal, Jennifer, Achim Oberg, and Justin J.W. Powell. "Das Verhältnis zwischen Hochschule und Wissenschaft in Deutschland: Expansion—Produktion—Kooperation." In *Komplexe Dynamiken globaler und lokaler Entwicklungen – 39. Kongress der Deutschen Gesellschaft für Soziologie in Göttingen*, ed. Deutsche Gesellschaft für Soziologie, 2019. https:// publikationen.soziologie.de/index.php/kongressband_2018/article/view/1109.

Dusdal, Jennifer, and Justin J.W. Powell. "Benefits, Motivations, and Challenges of International Collaborative Research: A Sociology of Science Case Study." *Science and Public Policy* 48, no. 2 (2021): 235–45. https://doi.org/10.1093/scipol/scab010.

Dusdal, Jennifer, Justin J.W. Powell, David P. Baker, Yuan Chih Fu, Yahya Shamekhi, and Manfred Stock. "University vs. Research Institute? The Dual Pillars of German Science Production, 1950–2010."*Minerva* 58, no. 3 (2020): 319–42. https://doi.org/10.1007/ s11024-019-09393-2.

Epping, Elizabeth. *Exploring the Institutionalisation of Science Diplomacy: A Comparison of German and Swiss Science and Innovation Centres.* Baden-Baden: Nomos Verlag, 2020.

European Commission. "Communication from the Commission to the Council, the European Parliament, the Economic and Social Committee and the Committee of the Regions: Social Policy Agenda." Last modified in 2000. https://eur-lex.europa.eu/LexUriServ/ LexUriServ.do?uri=CELEX:52000DC0379:EN:HTML (accessed May 21, 2022).

Evans, James A. "Industry Induces Academic Science to Know Less About More." *American Journal of Sociology* 116 (2010): 389–452. https://doi.org/10.1086/653834.

Fernandez, Frank, David P. Baker, Yuan Chih Fu, Ismael G. Muñoz, and Karly Sarita Ford. "A Symbiosis of Access: Proliferating STEM PhD Training in the US from 1920–2010." *Minerva* 59 (2021): 79–98. https://doi.org/10.1007/s11024-020-09422-5.

Flink, Tim. "Taking the Pulse of Science Diplomacy and Developing Practices of Valuation." *Science and Public Policy* 49 (2022): 191–200. https://doi.org/10.1093/scipol/scab074.

Frank, David John, and Jay Gabler. *Reconstructing the University: Worldwide Shifts in Academia in the 20th Century.* Stanford, CA: Stanford University Press, 2006.

Frank, David John, and John W. Meyer. *The University and the Global Knowledge Economy.* Princeton, NJ: Princeton University Press, 2020.

Franzen, Martina, Simone Rödder, and Peter Weingart. "Fraud: Causes and Culprits as Perceived by Science and the Media. Institutional Changes, Rather Than Individual Motivations, Encourage Misconduct." *EMBO Rep.* 8, no. 1 (2007): 3–7. doi: 10.1038/ sj.embor.7400884.

Fromherz, Allen J. *Qatar: A Modern History* (updated version). Washington, DC: Georgetown University Press, 2017.

Fu, Yuan Chih. "Science Production in Taiwanese Universities, 1980–2011." In *The Century of Science: The Global Triumph of the Research University,* ed. Justin J.W. Powell, David P. Baker, and Frank Fernandez, 173–204. Bingley, UK: Emerald, 2017. https:// doi.org/10.1108/S1479-367920170000033009.

Fu, Yuan Chih, David P. Baker, and Liang Zhang. "Engineering a World Class University? The Impact of Taiwan's World Class University Project on Scientific Productivity." *Higher Education Policy* 33, no. 3 (2020): 555–70. https://doi.org/10.1057/s41307-018-0110-z.

Fuller, Bruce, and Richard Rubinson. *The Political Construction of Education: The State, School Expansion, and Economic Change.* New York: Praeger, 1992.

Gauchat, Gordon. "The Cultural Authority of Science: Public Trust and Acceptance of Organized Science." *Public Understanding of Science* 20, no. 6 (2011): 751–70.

Gauffriau, Marianne, Peder Larsen, Isabelle Maye, Anne Roulin-Perriard, and Markus von Ins. "Comparisons of Results of Publication Counting Using Different Methods." *Scientometrics* 77, no. 1 (2008): 147–76. https://doi.org/10.1007/s11192-007-1934-2.

———. "Publication, Cooperation and Productivity Measures in Scientific Research." *Scientometrics* 73, no. 2 (2007): 175–214. https://doi.org/10.1007/s11192-007-1800-2.

Geiger, Roger L. *American Higher Education Since World War II.* Princeton, NJ: Princeton University Press, 2019.

———. *The History of American Higher Education: Learning and Culture from the Founding to World War II.* Princeton, NJ: Princeton University Press, 2016.

———, ed. *The Land-Grant Act and American Higher Education: Contexts and Consequences. History of Higher Education Annual.* New York: Routledge, 1998.

———. *Research and Relevant Knowledge: American Research Universities Since World War II.* Oxford, UK: Oxford University Press, 1993.

Geiger, Roger L., and Creso M. Sá. *Tapping the Riches of Science: Universities and the Promise of Economic Growth.* Cambridge, MA: Harvard University Press, 2008.

Georg-August-Universität Göttingen. "The History of the University." Georg-August-Universität Göttingen. https://www.uni-goettingen.de/en/history+of+the+university/52652.html (accessed May 12, 2022).

Gibbons, Michael, Camille Limoges, Helga Nowotny, Simon Schwartzman, Peter Scott, and Martin Trow. *The New Production of Knowledge: The Dynamics of Science and Research in Contemporary Societies.* London: SAGE, 1994.

Gillies, James. *CERN and the Higgs-Boson: The Global Quest for the Building Blocks of Reality.* London: Icon Books, 2018.

Gingras, Yves. *Bibliometrics and Research Evaluation: Uses and Abuses.* Cambridge, MA: MIT Press, 2016.

Glänzel, Wolfgang. "National Characteristics in International Scientific Co-Authorship Relations." *Scientometrics* 51, no. 1 (2001): 69–115. https://doi.org/10.1023/A:1010512628145.

Glänzel, Wolfgang. "Seven Myths in Bibliometrics About Facts and Fiction in Quantitative Science Studies." *COLLNET Journal of Scientometrics and Information Management* 2, no. 1 (2008): 9–17. https://doi.org/10.1080/09737766.2008.10700836.

Glänzel, Wolfgang, and András Schubert. "Analysing Scientific Networks Through Co-Authorship." In *Handbook of Quantitative Science and Technology Research*, ed. H. F. Moed, W. Glänzel, and U. Schmoch. Dordrecht, Netherlands: Springer, 2004. https://doi.org/10.1007/1-4020-2755-9_12.

Gläser, Jochen, and Werner Meske. *Anwendungsorientierung von Grundlagenforschung? Erfahrungen der Akademie der Wissenschaften der DDR.* Frankfurt am Main, Germany: Campus Verlag, 1996.

Gläser, Jochen, and Kathia Serrano-Velarde. "Changing Funding Arrangements and the Production of Scientific Knowledge: Introduction to the Special Issue." *Minerva* 56, no. 1 (2018): 1–10. https://doi.org/10.1007/s11024-018-9344-6.

Glass, Chris R., Stephanie Buus, and Larry A. Braskamp. *Uneven Experiences: What's Missing and What Matters for Today's International Students.* Chicago: Global Perspective Institute, 2013.

Godin, Benoit, and Yves Gingras. "The Place of Universities in the System of Knowledge Production." *Research Policy* 29, no. 2 (2000): 273–78. https://doi.org/10.1016/S0048-7333(99)00065-7.

Goldin, Claudia D., and Lawrence F. Katz. "The Race Between Education and Technology: The Evolution of US Educational Wage Differentials, 1890 to 2005." *NBER Working Paper Series* No. 12984. Cambridge, MA: National Bureau of Economic Research, 2007. https://doi.org/10.3386/w12984.

Goldin, Claudia D., and Lawrence F. Katz. *The Race Between Education and Technology.* Cambridge, MA: Harvard University Press, 2009.

Gongzhen, Li. "Historical Study of the University of Göttingen." *Chinese Studies in History* 50, no. 1 (2017): 38–50. https://doi.org/10.1080/00094633.2015.1189290.

Graf, Lukas, and Mathias Gardin. "Transnational Skills Development in Post-Industrial Knowledge Economies: The Case of Luxembourg and the Greater Region." *Journal of Education and Work* 31, no. 1 (2018): 1–15. https://doi.org/10.1080/13639080.2017.1408954.

Gui, Qinchang, Chengliang Liu, and Debin Du. "Globalization of Science and International Scientific Collaboration: A Network Perspective." *Geoforum* 105 (2019): 1–12. https://doi.org/10.1016/j.geoforum.2019.06.017.

Guston, David H., and Kenneth Keniston, eds. *The Fragile Contract: University Science and the Federal Government.* Cambridge, MA: MIT Press, 1994.

Halabi, Najeeb M., Alejandra Martinez, Halema Al-Farsi, Eliane Mery, Laurence Puydenus, Pascal Pujol, Hanif G. Khalak, et al. "Preferential Allele Expression Analysis Identifies Shared Germline and Somatic Driver Genes in Advanced Ovarian Cancer." *PLoS Genetics* 12, vol. 1 (2016): e1005755. https://10.1371/journal.pgen.1005755.

Harmsen, Robert, and Justin J.W. Powell. "Higher Education Systems and Institutions, Luxembourg." In *Encyclopedia of International Higher Education Systems and Institutions,* ed. Jung Cheol Shin and Pedro Teixeira. Heidelberg, Germany: Springer, 2018. https://doi.org/10.1007/978-94-017-9553-1_398-1.

Harzing, Anne-Wil, and Satu Alakangas. "Google Scholar, Scopus and the Web of Science: A Longitudinal and Cross-disciplinary Comparison." *Scientometrics* 106, no. 2 (2016): 787–804. https://doi.org/10.1007/s11192-015-1798-9.

Heim, Susanne, Carola Sachse, and Mark Walker, eds. *The Kaiser Wilhelm Society Under National Socialism.* Cambridge, UK: Cambridge University Press, 2009.

Heinze, Thomas, and Stefan Kuhlmann. "Across Institutional Boundaries? Research Collaboration in German Public Sector Nanoscience." *Research Policy* 37, no. 5 (2008): 888–99. doi: 10.1016/j.respol.2008.01.009.

Heinze, Thomas, Marie von der Heyden, and David Pithan. "Institutional Environments and Breakthroughs in Science: Comparison of France, Germany, the United Kingdom, and the United States." *PLoS One* 15, vol. 9 (2020): e0239805. https://doi.org/10.1371/journal.pone.0239805.

Helmich, Patricia, Sonia Gruber, and Rainer Frietsch. *Performance and Structures of the German Science System 2017. Studien zum deutschen Innovationssystem Nr. 5–2018.* Berlin: Expertenkommission Forschung und Innovation (EFI), 2018.

Henke, Justus, and Peer Pasternack. "Hochschulsystemfinanzierung. Wegweiser durch die Mittelströme." *HoF-Handreichungen* 9. Wittenberg, Germany: Institut für Hochschulforschung (HoF), 2017.

Hennicot-Schoepges, Erna. "Génèse d'un défi." In *Université du Luxembourg, 2003–2013,* ed. Michel Margue, 34–41. Luxembourg: Université du Luxembourg, 2013.

Hennicot-Schoepges, Erna. "Translation—Erna Hennicot-Schoepges." Interview by Samuel Hamen. *European Archive of Voices,* July 2020. https://arbeitaneuropa.com/transcripts/erna-hennicot-schoepges/ (accessed May 28, 2022).

Henrich, Joseph, Steven J. Heine, and Ara Norenzayan. "The Weirdest People in the World?" *Behavioral and Brain Sciences* 33, no. 2–3 (2010): 61–83. https://doi.org/10.1017/S0140525X0999152X.

Hessels, Laurens K., and Harro van Lente. "Re-Thinking New Knowledge Production: A Literature Review and a Research Agenda." *Research Policy* 37, no. 4 (2008): 740–60. https://doi.org/10.1016/j.respol.2008.01.008.

Hicks, Diana M., and J. Sylvan Katz. "Where Is Science Going?" *Science, Technology, & Human Values* 21, no. 4 (1996), 379–406. doi: 10.1177/016224399602100401.

Hinze, Sybille. "Forschungsförderung und ihre Finanzierung." In *Handbuch Wissenschaftspolitik*, ed. Dagmar Simon, Andreas Knie, Stefan Hornbostel, and Karin Zimmerman, 2nd ed., 413–28. Wiesbaden, Germany: VS Verlag für Sozialwissenschaften, 2016.

Hoelscher, Michael, and Julia Schubert. "Universities Between Inter- and Renationalization: An Introduction." *Global Perspectives* 3, no. 1 (2022): 56926. https://doi.org/10.1525/gp.2022.56926.

Hohn, Hans-Willy. "Governance-Strukturen und institutioneller Wandel des außeruniversitären Forschungssystems Deutschlands." In *Handbuch Wissenschaftspolitik*, ed. Dagmar Simon, Andreas Knie, Stefan Hornbostel, and Karin Zimmerman, 2nd ed., 549–72. Wiesbaden, Germany: VS Verlag für Sozialwissenschaften, 2016.

Hollingsworth, J. Rogers. "The Role of Institutions and Organizations in Shaping Radical Scientific Innovations." In *The Evolution of Path Dependence*, ed. Lars Magnusson and Jan Ottosson, 139–65. Cheltenham, UK: Edward Elgar, 2009.

Hollingsworth, J. Rogers, and Ellen J. Hollingsworth. *Major Discoveries, Creativity, and the Dynamics of Science*. Vienna: Remaprint, 2011.

Horgan, John. *The End of Science: Facing the Limits of Science in the Twilight of the Scientific Age*, 2nd ed. New York: Basic Books, 2015.

Horlacher, Rebecca. *The Educated Subject and the German Concept of Bildung: A Comparative Cultural History*. Abingdon, UK: Routledge, 2016.

Horta, Hugo. "Emerging and Near Future Challenges of Higher Education in East Asia." *Asian Economic Policy Review* 18 (2023): 171–91. https://doi.org/10.1111/aepr.12416.

Hottenrott, Hannah, and Cornelia Lawson. "A First Look at Multiple Institutional Affiliations: A Study of Authors in Germany, Japan and the UK." *Scientometrics* 111 (2017): 285–95. https://doi.org/10.1007/s11192-017-2257-6.

Huff, Toby E. *The Rise of Early Modern Science: Islam, China, and the West*. New York: Cambridge University Press, 2017.

Hunt, Michael H. "The American Remission of the Boxer Indemnity: A Reappraisal." *Journal of Asian Studies* 31, no. 3 (1972): 539–59. https://www.jstor.org/stable/2052233.

Hüther, Otto, and Georg Krücken. *Higher Education in Germany—Recent Developments in an International Perspective*. Cham, Switzerland: Springer, 2018.

IceCube Collaboration: Rasha Abbasi, Markus Ackermann, et al. "Detection of Astrophysical Tau Neutrino Candidates in IceCube." *The European Physical Journal* C 82, no. 1031 (2022). https://10.1140/epjc/s10052-022-10795-y.

IceCube Collaboration. "Evidence for Neutrino Emission from the Nearby Active Galaxy NGC 1068." *Science* 378, no. 6619 (2022): 538–43.arXiv:2211.09972.

IceCube Neutrino Observatory. "IceCube." https://icecube.wisc.edu/. (accessed May 25, 2022).

Iggers, Georg G. "The Intellectual Foundations of Nineteenth-Century 'Scientific' History: The German Model." In *Oxford History of Historical Writing, vol. 4: 1800–1945*, ed. Stuart Macintyre, Juan Maiguashca, and Attila Pók. Oxford, UK: Oxford University Press, 2011. https://doi.org/10.1093/acprof:osobl/9780199533091.003.0003.

Jarausch, Konrad. "American Students in Germany, 1815–1914: The Structure of German and U.S. Matriculants at Göttingen University." In *German Influences on Education in the*

United States to 1917, ed. H. Geitz, J. Heideking, and J. Herbst, 195–212. Cambridge, UK: Cambridge University Press, 1995.

Jones, Benjamin F., Stefan Wuchty, and Brian Uzzi. "Multi-University Research Teams: Shifting Impact, Geography, and Stratification in Science." *Science* 322, no. 5905 (2008): 1259–62. doi: 10.1126/science.1158357.5905.

Jöns, Heike. "Transnational Mobility and the Spaces of Knowledge Production: A Comparison of Global Patterns, Motivations and Collaborations in Different Academic Fields." *Social Geography* 2, no. 2 (2007): 97–114. https://doi.org/10.5194/sg-2-97-2007.

Jöns, Heike, Peter Meusburger, and Michael Heffernan, eds. *Mobilities of Knowledge.* Cham, Switzerland: Springer Nature, 2017. https://doi.org/10.1007/978-3-319-44654-7.

Kamola, Isaac A. *Making the World Global. U.S. Universities and the Production of the Global Imaginary.* Durham, NC: Duke University Press, 2019.

Kennedy, Michael D. *Globalizing Knowledge: Intellectuals, Universities, and Publics in Transformation.* Stanford, CA: Stanford University Press, 2014.

Kerr, Clark. *The Uses of the University.* 5th ed. Cambridge, MA: Harvard University Press, [1963] 2001.

Kett, Joseph F. *The Pursuit of Knowledge Under Difficulties: From Self-improvement to Adult Education in America, 1750–1990.* Stanford, CA: Stanford University Press, 1994.

Kienast, Sarah-Rebecca. "How Do Universities' Organizational Characteristics, Management Strategies, and Culture Influence Academic Research Collaboration? A Literature Review and Research Agenda." *Tertiary Education and Management* (2023). https://doi.org/10.1007/s11233-022-09101-y.

Kim, Hyerim, and Junghee Choi. "The Growth of Higher Education and Science Production in South Korea Since 1945." In *The Century of Science: The Global Triumph of the Research University*, ed. Justin J.W. Powell, David P. Baker, and Frank Fernandez, 205–26. Bingley, UK: Emerald, 2017. https://doi.org/10.1108/S1479-367920170000033010.

Kirby, William C. *Empires of Ideas: Creating the Modern University from Germany to America to China.* Cambridge, MA: Belknap Press of Harvard University Press, 2022.

Kirby, William C., and Marijk van der Wende. "The New Silk Road: Implications for Higher Education in China and the West?" *Cambridge Journal of Regions, Economy and Society* 12, no. 1 (2019): 127–44. https://doi.org/10.1093/cjres/rsy034.

Klemenčič, Manja. "Epilogue: Reflections on a New Flagship University." In *The New Flagship University*, ed. John Aubrey Douglass, 191–97. Basingstoke, UK: Palgrave Macmillan, 2016.

Kmiotek-Meier, Emilia, Ute Karl, and Justin J.W. Powell. "Designing the (Most) Mobile University: The Centrality of International Student Mobility in Luxembourg's Higher Education Policy Discourse." *Higher Education Policy* 33, vol. 1 (2020): 21–44. https://doi.org/10.1057/s41307-018-0118-4.

Kmiotek-Meier, Emilia, and Justin J.W. Powell. "Evaluating Universal Student Mobility: Contrasting Policy Discourse and Student Narratives in Luxembourg." *International Studies in Sociology of Education* 32, no. 2 (2022): 466–86. doi: 10.1080/09620214.2021.2007416.

Kosmützky, Anna. "A Two-Sided Medal. On the Complexities of Collaborative and Comparative Team Research." *Higher Education Quarterly* 72, no. 4 (2018): 314–31. https://doi.org/10.1111/hequ.12156.

Kosmützky, Anna, and Georg Krücken. "Governing Research. New Forms of Competition and Cooperation in German Academia." In *Restoring Collegiality: Revitalizing Faculty*

Authority in Universities, ed. Kerstin Sahlin and Ulla Eriksson-Zetterquist, 31–57. Bingley, UK: Emerald, 2023.

———. "Still the Century of the University as a Global Institution? Comparative Perspectives." *Global Perspectives* 4, no. 1 (2023): 68084. https://doi.org/10.1525/gp.2023.68084.

Kosmützky, Anna, and Romy Wöhlert. "Varieties of Collaboration: On the Influence of Funding Schemes on Forms and Characteristics of International Collaborative Research Projects (ICRPs)." *European Journal of Education* 56, no. 2 (2022): 182–99. https://doi.org/10.1111/ejed.12452.

Kozlov, Max. "'Disruptive' Science Has Declined—and No One Knows Why. *Nature* 613, no. 225 (2023). https://doi.org/10.1038/d41586-022-04577-5.

Krücken, Georg. "Multiple Competitions in Higher Education: A Conceptual Approach." *Innovation* 23, no. 2 (2021): 163–81. https://doi.org/10.1080/14479338.2019.1684652.

Krüger, Anne K. "Quantification 2.0? Bibliometric Infrastructures in Academic Evaluation." *Politics and Governance* 8, no. 2 (2020): 58–67. https://doi.org/10.17645/pag.v8i2.2575.

Kuhn, Thomas S. *The Structure of Scientific Revolutions*. 50th Anniversary ed. Chicago: University of Chicago Press, [1962] 2012.

Kwiek, Marek. "What Large-Scale Publication and Citation Data Tell Us About International Research Collaboration in Europe: Changing National Patterns in Global Contexts." *Studies in Higher Education* 46, no. 12 (2021): 2629–49. https://doi.org/10.1080/03075079.2020.1749254.

Labaree, David F. *A Perfect Mess: The Unlikely Ascendancy of American Higher Education*. Chicago: University of Chicago Press, 2017.

Ladyman, James. *Understanding Philosophy of Science*. London: Routledge, 2001. https://doi.org/10.4324/9780203463680.

Larivière, Vincent, Yves Gingras, Cassidy R. Sugimoto, and Andrew Tsou. "Team Size Matters: Collaboration and Scientific Impact Since 1900." *Journal of the Association for Information Science and Technology* 66, no. 7 (2015): 1323–32. https://doi.org/10.1002/asi.23266.

Larivière, Vincent, Chaoqun Ni, Yves Gingras, Blaise Cronin, and Cassidy R. Sugimoto. "Bibliometrics: Global Gender Disparities in Science." *Nature News* 504, no. 7479 (2013): 211. https://doi.org/10.1038/504211a.

Laudel, Grit. "What Do We Measure by Co-Authorships?" *Research Evaluation* 11, no. 1 (2002): 3–15. https://doi.org/10.3152/147154402781776961.

Lee, Jack T. "Education Hubs and Talent Development: Policymaking and Implementation Challenges." *Higher Education* 68 (2014): 807–23. https://doi.org/10.1007/s10734-014-9745-x.

Leibfried, Stefan, ed. *Die Exzellenzinitiative. Zwischenbilanz und Perspektiven*. Frankfurt am Main, Germany: Campus Verlag, 2010.

Lenhardt, Gero. *Hochschulen in Deutschland und den USA*. Wiesbaden, Germany: VS Verlag für Sozialwissenschaften, 2005.

Leslie, Stuart W. *The Cold War and American Science: The Military-Industrial-Academic Complex at MIT and Stanford*. New York: Columbia University Press, 1993.

Levine, Emily J. *Allies and Rivals: German-American Exchange and the Rise of the Modern Research University*. Chicago: University of Chicago Press, 2021.

———. "Baltimore Teaches, Göttingen Learns: Cooperation, Competition, and the Research University." *The American Historical Review* 121 (2016): 780–823. https://doi.org/10.1093/ahr/121.3.780.

Leydesdorff, Loet, Henry Etzkowitz, and Duncan Kushnir. "Globalization and Growth of US University Patenting (2009–2014)." *Industry and Higher Education* 30, no. 4 (2016): 257–66. https://doi.org/10.1177/0950422216660253.

Leydesdorff, Loet, and Caroline Wagner. "Is the United States Losing Ground in Science? A Global Perspective on the World Science System." *Scientometrics* 78, no. 1 (2009): 23–36. https://doi.org/10.1007/s11192-008-1830-4.

LIGO. "LIGO Lab." https://www.ligo.caltech.edu/ (accessed May 25, 2022).

Lovakov, Andrey, Maia Chankseliani, and Anna Panova. "Universities vs. Research Institutes? Overcoming the Soviet Legacy of Higher Education and Research." *Scientometrics* 127 (2022): 6293–6313. https://doi.org/10.1007/s11192-022-04527-y.

Mann, Michael. *The Sources of Social Power, Volume 1: A History of Power from the Beginning to AD 1760*. New York: Cambridge University Press, 2012.

Marginson, Simon. "'All Things Are in Flux': China in Global Science." *Higher Education* 83 (2022): 881–910. https://doi-org.proxy.bnl.lu/10.1007/s10734-021-00712-9.

Marginson, Simon, and Xin Xu, eds. *Changing Higher Education in East Asia*. London: Bloomsbury, 2022.

Margue, Michel, ed. *Université du Luxembourg 2003–2013*. Luxembourg: University of Luxembourg, 2013.

Marques, Marcelo. "Governing European Educational Research Through Ideas? Incremental Ideational Change in the European Union's Framework Programme (1994–2020)." *European Journal of Education* (2023). doi: 10.1111/ejed.12579.

Marques, Marcelo, and Lukas Graf. "Pushing Boundaries: The European Universities Initiative as a Case of Transnational Institution Building."*Minerva* (2023). doi: 10.1007/s11024-023-09516-w.

Marques, Marcelo, Mike Zapp, and Justin J.W. Powell. "Europeanizing Universities: Expanding and Consolidating Networks of the Erasmus Mundus Joint Master Degree Programme (2004–2017)." *Higher Education Policy* 35 (2022): 19–41. https://doi.org/10.1057/s41307-020-00192-z.

Matthies, Annemarie, and Manfred Stock. "Universitätsstudium und berufliches Handeln. Eine historisch-soziologische Skizze zur Entstehung des 'Theorie-Praxis-Problems.'" In *Wieviel Wissenschaft braucht die Lehrerbildung?* ed. Claudia Scheid and Thomas Wenzl, 215–53. Wiesbaden, Germany: Springer VS, 2020.

May, Robert M. "The Scientific Wealth of Nations." *Science* 275 (1997): 793–96.

Mayntz, Renate. "Academy of Sciences in Crisis: A Case Study of a Fruitless Struggle for Survival." In *Coping with Trouble: How Science Reacts to Political Disturbances of Research Conditions*, ed. Uwe Schimank and Andreas Stucke, 163–88. Frankfurt am Main, Germany: Campus Verlag, 1994.

Max-Planck-Gesellschaft. "Adolf Harnack's Memorandum for a Reform of German Science." 2022. https://www.mpg.de/947056/1_person0-1909 (accessed June 2, 2022).

McClelland, Thomas E. *State, Society, and University in Germany: 1700–1914*. Cambridge, UK: Cambridge University Press, 1980.

McCurry, Justin. "Disgrace." *The Guardian*, January 1, 2006.

Mehta, Jal, and Scott Davies, eds. *Education in a New Society: Renewing the Sociology of Education*. Chicago: University of Chicago Press, 2018.

Menand, Louis, Paul Reitter, and Chad Wellmon. *The Rise of the Modern Research University: A Sourcebook*. Chicago: University of Chicago Press, 2017.

Merton, Robert K. "The Matthew Effect in Science, II: Cumulative Advantage and the

Symbolism of Intellectual Property." *Isis* 79, no. 4 (1988): 606–23. https://www.jstor.org/stable/234750.

———. *The Sociology of Science: Theoretical and Empirical Investigations.* Chicago: University of Chicago Press, 1973.

Merton, Robert K., and Elinor Barber. *The Travels and Adventures of Serendipity: A Study in Sociological Semantics and the Sociology of Science.* Princeton, NJ: Princeton University Press, 2006.

Meyer, Heinz-Dieter. *The Design of the University: German, American, and "World Class."* New York: Routledge, 2017.

Meyer, Martin. "Does Science Push Technology? Patents Citing Scientific Literature." *Research Policy* 29, no. 3 (2000): 409–34. https://doi.org/10.1016/S0048-7333(99)00040-2.

Meyer, Morgan B. "Creativity and Its Contexts: The Emergence, Institutionalisation and Professionalisation of Science and Culture in Luxembourg." *European Review of History/Revue Europeenne d'Histoire* 16, no. 4 (2009): 453–76. https://doi.org/10.1080/13507480903063605.

———. "The Dynamics of Science in a Small Country: The Case of Luxembourg." *Science and Public Policy* 35, no. 5 (2008): 361–71. https://doi.org/10.3152/030234208X317133.

Miao, Lili, Dakota Murray, Woo-Sung Jung, Vincent Larivière, Cassidy R. Sugimoto, and Yong-Yeol Ahn. "The Latent Structure of Global Scientific Development." *Nature Human Behavior* 6, no. 9 (2022): 1206–17. doi: 10.1038/s41562-022-01367-x.

Minerva Jahrbuch der gelehrten Welt: Abteilung Universitäten und Fachhochschulen. Berlin: de Gruyter. https://www.degruyter.com/serial/minervjb-b/html?lang=en (accessed May 28, 2022).

Mitterle, Alexander, and Manfred Stock. "Higher Education Expansion in Germany: Between Civil Rights, State-Organized Entitlement System and Academization." *European Journal of Higher Education* 11, no. 3 (2021): 292–311. https://doi.org/10.1080/21568235.2021.1944815.

Mohrman, Kathryn, Wanhua Ma, and David P. Baker. "The Research University in Transition: The Emerging Global Model." *Higher Education Policy* 21, no. 1 (2008): 5–27. https://doi.org/10.1057/palgrave.hep.8300175.

Mosbah-Natanson, Sébastien, and Yves Gingras. "The Globalization of Social Sciences?" *Current Sociology* 62, no. 5 (2013): 626–46. https://doi.org/10.1177/0011392113498866.

Münch, Richard. *Die akademische Elite: Zur sozialen Konstruktion wissenschaftlicher Exzellenz.* Frankfurt am Main, Germany: Suhrkamp, 2007.

Murakami, Yoichiro P. "Scientization of Science." *Annals of the Japan Association for Philosophy of Science* 8, no. 3 (1993): 175–85.

———. "Transformation of Science." *Annals of the Japan Association for Philosophy of Science* 9, no. 2 (1997): 79–85.

Musselin, Christine. "Bringing Universities to the Centre of the French Higher Education System? Almost But Not Yet. . . ." *European Journal of Higher Education* 11, no. 3 (2021): 329–45. https://doi.org/10.1080/21568235.2021.1945474.

———. *La Grande Course des Universités.* Paris: Presse de Sciences Po, 2017.

———. "New Forms of Competition in Higher Education." *Socio-Economic Review* 16, no. 3 (2018): 657–83. https://doi.org/10.1093/ser/mwy033.

Musser, George. "A Defense of the Reality of Time." *Quanta Magazine,* May 16, 2017. https://www.quantamagazine.org/a-defense-of-the-reality-of-time-20170516/.

Nam, Seung Wan. "An Assessment of the Impact of the Center of Excellence Program on the Research Production of Korean Universities from 1989 to 2011." PhD dissertation, Pennsylvania State University, 2020.

Narin, Francis, Kimberly S. Hamilton, and Dominic Olivastro. "The Increasing Linkage Between US Technology and Public Science." *Research Policy* 26, no. 3 (1997): 317–30. https://doi.org/10.1016/S0048-7333(97)00013-9.

National Human Genome Research Institute. "What Is the Human Genome Project?" https://www.genome.gov/human-genome-project/What (accessed June 30, 2021).

National Science Board, National Science Foundation. "Publications Output: U.S. and International Comparisons." Science and Engineering Indicators 2022. NSB-2021-4. Alexandria, VA. 2021. https://ncses.nsf.gov/pubs/nsb20214.

National Science Board, National Science Foundation. "The State of U.S. Science and Engineering 2020." https://ncses.nsf.gov/pubs/nsb20201 (accessed May 25, 2022).

National Science Foundation. *Doctorate Recipients from U.S. Universities.* National Center for Science and Engineering Statistics, Directorate for Social, Behavioral and Economic Sciences, 2017. https://www.nsf.gov/statistics/2017/nsf17306/static/report/nsf17306.pdf.

Ness, Roberta B. *The Creativity Crisis: Reinventing Science to Unleash Possibility.* Oxford, UK: Oxford University Press, 2015.

"Neutrinos Build a Ghostly Map of the Milky Way: Astronomers for the First Time Detected Neutrinos That Originated Within Our Local Galaxy Using a New Technique," *The New York Times*, June 29, 2023, https://www.nytimes.com/2023/06/29/science/neutrinos-milky-way-map.html.

Novikoff, Alex J. *The Medieval Culture of Disputation: Pedagogy, Practice, and Performance.* Philadelphia: University of Pennsylvania Press, 2013.

Nowotny, Helga, Peter Scott, and Michael Gibbons. "'Mode 2' Revisited: The New Production of Knowledge." *Minerva* 41 (2003): 179–94. https://doi.org/10.1023/A:1025505528250.

OECD. "OECD.stat. 2019, Main Science and Technology Indicators." https://stats.oecd.org/Index.aspx?DataSetCode=MSTI_PUB. (accessed July 25, 2022).

OECD Indicators. "Education at a Glance 2022, with a Spotlight on Tertiary Education." Paris: Organisation for Economic Co-operation and Development, 2022. https://www.oecd.org/education/education-at-a-glance.

Okamura, Keisuke. "A Half-Century of Global Collaboration in Science and the 'Shrinking World'." *Quantitative Science Studies* (2023). doi: doi.org/10.1162/qss_a_00268.

Oleksiyenko, Anatoly. "On the Shoulders of Giants? Global Science, Resource Asymmetries, and Repositioning of Research Universities in China and Russia." *Comparative Education Review* 58, no. 3 (2016): 482–508. https://doi.org/10.1086/676328.

Ortiga, Yasmin Y., Meng-Hsuan Chou, and Jue Wang. "Competing for Academic Labor: Research and Recruitment Outside the Academic Center." *Minerva* 58 (2020): 607–24. https://doi-org.proxy.bnl.lu/10.1007/s11024-020-09412-7.

Owen-Smith, Jason. *Research Universities and the Public Good: Discovery for an Uncertain Future.* Stanford, CA: Stanford Business Books, 2018.

Park, Michael, Erin Leahey, and Russell J. Funk. "Papers and Patents Are Becoming Less Disruptive Over Time." *Nature* 613 (2023): 138–44. https://doi.org/10.1038/s41586-022-05543-x.

Parsons, Talcott, and Gerald M. Platt. *The American University.* Cambridge, MA: Harvard University Press, 1973.

Pauck, Wilhelm. *Harnack and Troeltsch: Two Historical Theologians.* Eugene, OR: Wipf and Stock, 2015.

Peacock, Vita. "Academic Precarity as Hierarchical Dependence in the Max Planck Society." *HAU: Journal of Ethnographic Theory* 6, no. 1 (2016): 95–119. https://doi.org/10.14318/hau6.1.006.

Posselt, Julie R., Ozan Jaquette, Rob Bielby, and Michael N. Bastedo. "Access Without Equity: Longitudinal Analyses of Institutional Stratification by Race and Ethnicity, 1972–2004." *American Educational Research Journal* 49, no. 6 (2012): 1074–1111. https://doi.org/10.3102/0002831212439456.

Powell, Justin J.W. *Barriers to Inclusion: Special Education in the United States and Germany.* Abingdon: Routledge, [2011] 2016.

———. "Higher Education and the Exponential Rise of Science: Competition and Collaboration." In *Emerging Trends in the Social and Behavioral Sciences*, ed. Robert Scott and Marlis Buchmann. Hoboken, NJ: John Wiley & Sons, 2018. https://onlinelibrary.wiley.com/doi/book/10.1002/9781118900772.

———. "International National Universities: Migration and Mobility in Luxembourg and Qatar." In *Internationalisation of Higher Education and Global Mobility*, ed. Bernhard Streitwieser, 119–33. Oxford: Symposium Books, 2014.

Powell, Justin J.W., David P. Baker, and Frank Fernandez, eds. *The Century of Science: The Global Triumph of the Research University.* International Perspectives on Education and Society, vol. 33. Bingley, UK: Emerald, 2017.

Powell, Justin J.W., Nadine Bernhard, and Lukas Graf. "The Emergent European Model in Skill Formation: Comparing Higher Education and Vocational Training in the Bologna and Copenhagen Processes." *Sociology of Education* 85, no. 3 (2012): 240–58. https://doi.org/10.1177/0038040711427313.

Powell, Justin J.W., and Jennifer Dusdal. "The European Center of Science Productivity: Research Universities and Institutes in France, Germany, and the United Kingdom." In *The Century of Science: The Global Triumph of the Research University*, ed. Justin J.W. Powell, David P. Baker, and Frank Fernandez, 55–84. Bingley, UK: Emerald, 2017. https://doi.org/10.1108/S1479-367920170000033005.

———. "Science Production in Germany, France, Belgium, and Luxembourg: Comparing the Contributions of Research Universities and Institutes to Science, Technology, Engineering, Mathematics, and Health." *Minerva* 55 (2017): 413–34. https://doi.org/10.1007/s11024-017-9327-z.

Powell, Justin J.W., Frank Fernandez, John T. Crist, Jennifer Dusdal, Liang Zhang, and David P. Baker. "Introduction: The Worldwide Triumph of the Research University and Globalizing Science." In *The Century of Science: The Global Triumph of the Research University*, ed. Justin J.W. Powell, David P. Baker, and Frank Fernandez, 1–36. Bingley, UK: Emerald, 2017. https://doi.org/10.1108/S1479-367920170000033003.

Powell, Justin J.W., and Heike Solga. "Why Are Higher Education Participation Rates in Germany So Low? Institutional Barriers to Higher Education Expansion." *Journal of Education and Work* 24, vol. 1 (2011): 49–68. https://doi.org/10.1080/13639080.2010.534445.

Powell, Walter W. "Learning from Collaboration," *California Management Review* 40, no. 3 (1998): 228–40. https://doi.org/10.2307/41165952.

Powell, Walter W., Kenneth W. Koput, and Laurel Smith-Doerr. "Interorganizational Collaboration and the Locus of Innovation: Networks of Learning in Biotechnology." *Administrative Science Quarterly* 41 (1996): 116–45. https://doi.org/10.2307/2393988.

Powell, Walter W., and Kaisa Snellman. "The Knowledge Economy." *Annual Review of Sociology* 30 (2004): 199–220. https://doi.org/10.1146/annurev.soc.29.010202.100037.

Price, Derek J. de Solla. *Little Science, Big Science.* New York: Columbia University Press, 1963.

———. *Little Science, Big Science . . . and Beyond.* New York: Columbia University Press, 1986.

———. "Networks of Scientific Papers." *Science* 149, no. 3683 (1965): 510–15.

———. "The Science of Science." *Bulletin of the Atomic Scientists* 21, no. 8 (1965): 2–8.

Pritchard, Rosalind. "Trends in the Restructuring of German Universities." *Comparative Education Review* 50, no. 1 (2006): 90–112. https://doi.org/10.1086/498330.

Psillos, Stathis. *Scientific Realism: How Science Tracks Truth.* London: Routledge, 1999.

Qatar Foundation. "About Qatar Foundation." https://www.qf.org.qa/about. (accessed July 9, 2021)

Rawlings, Craig M., and Michael D. Bourgeois. "The Complexity of Institutional Niches: Credentials and Organizational Differentiation in a Field of US Higher Education." *Poetics* 32, no. 6 (2004): 411–46. https://doi.org/10.1016/S0304-422X(04)00057-9.

Reid, Constance. *Hilbert-Courant.* Berlin: Springer, 1986.

Reisz, Robert D. "Göttingen in Baltimore or the Americanization of the German University?" *Journal of Research in Higher Education* 2, no. 2 (2018): 23–44. http://dx.doi.org/10.24193/JRHE.2018.2.2.

Retraction Watch. "Retraction Watch: Tracking Retractions as a Window into the Scientific Process." https://retractionwatch.com/ (accessed May 11, 2022).

Richardson, John G., and Justin J.W. Powell. *Comparing Special Education: Origins to Contemporary Paradoxes.* Stanford, CA: Stanford University Press, 2011.

Riesman, David, Nathan Glazer, and Reuel Denney. *The Lonely Crowd: A Study of the Changing American Character.* New Haven, CT: Yale University Press, 1950.

Ringer, Fritz K. *The Decline of the German Mandarins: The German Academic Community, 1890–1933.* Middletown, CT: Wesleyan University Press, 1990.

Robin, Stéphane, and Torben Schubert. "Cooperation with Public Research Institutions and Success in Innovation: Evidence from France and Germany." *Research Policy* 42, no. 1 (2013): 149–66. https://doi.org/10.1016/j.respol.2012.06.002.

Rohrbeck, René. "F+E-Politik von Unternehmen." In *Handbuch Wissenschaftspolitik,* ed. Dagmar Simon, Andreas Knie, and Stefan Hornbostel, 427–40. Wiesbaden: VS Verlag für Sozialwissenschaften, 2010.

Rohstock, Anne. "Wider die Gleichmacherei! Luxemburgs langer Weg zur Universität 1848–2003." *forum* 301 (2010): 43–46.

Rohstock, Anne, and Catherina Schreiber. "The Grand Duchy on the Grand Tour: A Historical Study of Student Migration in Luxembourg." *Paedagogica Historica* 49, no. 2 (2013): 174–93. https://doi.org/10.1080/00309230.2012.701221.

Rothblatt, Sheldon, ed. *Clark Kerr's World of Higher Education Reaches the 21st Century: Chapters in a Special History.* Heidelberg: Springer, 2012. https://doi.org/10.1007/978-94-007-4258-1.

The Royal Society. *Knowledge, Networks and Nations: Global Scientific Collaboration in the 21st Century.* London: The Royal Society, 2011. https://royalsociety.org/~/media/royal_society_content/policy/publications/2011/4294976134.pdf.

Rüegg, Walter. *Vom 19. Jahrhundert zum Zweiten Weltkrieg 1800–1945. Geschichte der Universität in Europa,* vol. III. München: C.H. Beck, 2004.

Salsburg, David. *The Lady Tasting Tea: How Statistics Revolutionized Science in the Twentieth Century.* New York: W.H. Freeman, 2001.

Schaub, Maryellen, Hyerim Kim, Deok-ho Jang, and David P. Baker. "Policy Reformer's Dream or Nightmare?" *Compare: A Journal of Comparative and International Education* 50, no. 7 (2020): 1066–79. https://doi.org/10.1080/03057925.2020.1733797.

Schofer, Evan, and John W. Meyer. "The Worldwide Expansion of Higher Education in the Twentieth Century." *American Sociological Review* 70, no. 6 (2005): 898–920. https://doi.org/10.1177/000312240507000602.

Schofer, Evan, Francisco O. Ramirez, and John W. Meyer. "The Societal Consequences of Higher Education." *Sociology of Education*, 94, no. 1 (2021): 1–19. https://doi.org/10.1177/0038040720942912.

Schofer, Evan, Julia C. Lerch, and John W. Meyer, "Illiberal Reactions to Higher Education."*Minerva* 60 (2022): 509–34. doi: 10.1007/s11024-022-09472-x.

Shapin, Steven. *The Scientific Revolution.* Chicago: University of Chicago Press, 2018.

Shils, Edward. *Max Weber on Universities: The Power of the State and the Dignity of the Academic Calling in Imperial Germany.* Chicago: University of Chicago Press, 1973.

Shima, Kazunori. "Changing Science Production in Japan: The Expansion of Competitive Funds, Reduction of Block Grants, and Unsung Heroes." In *The Century of Science: The Global Triumph of the Research University*, ed. Justin J.W. Powell, David P. Baker, and Frank Fernandez, 113–40. Bingley, UK: Emerald, 2017. https://doi.org/10.1108/S1479-367920170000033007.

Smelser, Neil. *Dynamics of the Contemporary University: Growth, Accretion, and Conflict.* Berkeley: University of California Press, 2013. https://doi.org/10.1525/9780520955257.

Smith, Bruce L. R. *American Science Policy Since World War II.* Washington, DC: Brookings Institution Press, 1990.

STATEC. "626,000 Inhabitants as of January 1, 2020." Press release, January 4, 2020. https://statistiques.public.lu/en/news/population/2020/04/20200401/index.html. (accessed February 11, 2021).

Stephan, Paula E. *How Economics Shapes Science.* Cambridge, MA: Harvard University Press, 2012.

Stevens, Mitchell L., Elizabeth Armstrong, and Richard Arum. "Sieve, Incubator, Temple, Hub: Empirical and Theoretical Advances in the Sociology of Higher Education." *Annual Review of Sociology* 34 (2008): 127–51. https://doi.org/10.1146/annurev.soc.34.040507.134737.

Stevens, Mitchell L., Cynthia Miller-Idriss, and Seteney Shami. *Seeing the World: How US Universities Make Knowledge in a Global Era.* Princeton, NJ: Princeton University Press, 2018.

Stevenson, Louise L. "Scholarly Means to Evangelical Ends: The New Haven Scholars, 1840–1890." PhD dissertation, Boston University, 1981.

Stichweh, Rudolf. "The University as a World Organization." 2022. https://www.researchgate.net/publication/359046520_The_University_as_a_World_Organization.

———. *Wissenschaft, Universität, Professionen.* Frankfurt am Main, Germany: Suhrkamp, 1994.

———. *Wissenschaft, Universität, Professionen*, 2nd ed. Bielefeld. Germany: Transcript, 2013.

Stigler, George J. "The Intellectual and the Marketplace." In *The Intellectual and the Marketplace*, 143–58. Cambridge, MA: Harvard University Press, 2014.

Stock, Manfred. "Hochschulexpansion und Akademisierung der Beschäftigung." *Soziale Welt* 68, no. 4 (2018): 347–64. 10.5771/0038-6073-2017-4-347.

Stock, Manfred, Justin J.W. Powell, and Robert D. Reisz. "Higher Education and Scientific

Research in Germany: Reconstructing the Nexus of Research and Teaching." In *Education and Society: In Memoriam Robert Reisz*, ed. Silviu E. Rogobete and Emanuel Copilaș, 157–83. Timișoara, Romania: Editura Universității de Vest, 2023.

Sugimoto, Cassidy R., and Vincent Larivière. *Measuring Research: What Everyone Needs to Know*. Oxford, UK: Oxford University Press, 2018.

Sugimoto, Cassidy R., Nicolás Robinson-García, Dakota S. Murray, Alfredo Yegros-Yegros, Rodrigo Costas, and Vincent Larivière. "Scientists Have Most Impact When They're Free to Move." *Nature News* 550, no. 7674 (2017): 29–31.

Tahamtan, Iman, and Lutz Bornmann. "Altmetrics and Societal Impact Measurements: Match or Mismatch? A Literature Review." *El Profesional de la Información* 29, no. 1 (2020): e290102. https://doi.org/10.3145/epi.2020.ene.02.

Tenopir, Carol, and Donald W. King. "The Growth of Journals Publishing." In *The Future of the Academic Journal*, ed. Bill Cope and Angus Phillips, 159–78. Witney, UK: Chandos, 2014.

Torka, Marc. *Die Projektförmigkeit der Forschung*. Baden-Baden, Germany: Nomos, 2009.

Treviño, A. Javier. *Talcott Parsons Today: His Theory and Legacy in Contemporary Sociology*. Lanham, MD: Rowman & Littlefield, 2001.

UNESCO. "UIS Releases New Data for SDG 9.5 on Research and Development." May 6, 2022. https://uis.unesco.org/en/news/uis-releases-new-data-sdg-9-5-research-and-development. (accessed July 26, 2022).

Urquiola, Miguel S. *Markets, Minds, and Money: Why America Leads the World in University Research*. Cambridge, MA: Harvard University Press, 2020.

Vanderstraeten, Raf. "The Making of Parsons's *The American University*." *Minerva* 53, no. 4 (2015): 307–25. https://doi.org/10.1007/s11024-015-9285-2.

Vierhaus, Rudolf. "Bemerkungen zum sogenannten Harnack-Prinzip Mythos und Realität." In *Die Kaiser–Wilhelm/Max Planck Gesellschaft und ihre Institute*, ed. Bernhard vom Brocke and Hubert Laitko, 129–38. Berlin: de Gruyter, 1996.

Viglione, Giuliana. "China Is Closing the Gap with the United States on Research Spending." *Nature News*, January 15, 2020. https://doi.org/10.1038/d41586-020-00084-7.

vom Brocke, Bernhard. "Wege aus der Krise: Universitätsseminar, Akademiekommission oder Forschungsinstitut. Formen der Institutionalisierung in den Geistes- und Naturwissenschaften 1810–1900–1995." In *Konkurrenten in der Fakultät. Kultur, Wissen und Universität um 1900*, ed. Christoph König and Eberhard Lämmert, 191–217. Frankfurt am Main, Germany: Fischer, 1999.

Wagner, Caroline S. *The Collaborative Era in Science: Governing the Network*. Basingstoke, UK: Palgrave Macmillan, 2018.

Waldinger, Fabian. "Peer Effects in Science: Evidence from the Dismissal of Scientists in Nazi Germany." *The Review of Economic Studies* 79, no. 2 (2011): 838–61. https://doi.org/10.1093/restud/rdr029.

Wang, Dashun, and Albert-László Barabási. *The Science of Science*. Cambridge, UK: Cambridge University Press, 2021.

Wang, Lucy Lu, Kyle Lo, Yoganand Chandrasekhar, Russell Reas, Jiangjiang Yang, Darrin Eide, Kathryn Funk, et al. "Cord-19: The COVID-19 Open Research Dataset." Preprint, submitted April 22, 2020. https://doi.org/10.48550/arXiv.2004.10706.

Watson, Peter. *The German Genius: Europe's Third Renaissance, the Second Scientific Revolution and the Twentieth Century*. New York: HarperCollins, 2010.

Web of Science Group. "*Web of Science* Core Collection." Last modified May 19, 2022. https://mjl.clarivate.com/collection-list-downloads. (accessed May 20, 2022).

Weingart, Peter. "Growth, Differentiation, Expansion and Change of Identity—The Future of Science." In *Social Studies of Science and Technology: Looking Back, Ahead* (Sociology of Sciences Yearbook), ed. Bernward Joerges and Helga Nowotny, 183–200. Dordrecht, NL: Kluwer, 2003.

Weingart, Peter, and Matthias Winterhager. *Die Vermessung der Forschung: Theorie und Praxis der Wissenschaftsindikatoren.* Frankfurt am Main, Germany: Campus Verlag, 1984.

Whitley, Richard, Jochen Gläser, and Grit Laudel. "The Impact of Changing Funding and Authority Relationships on Scientific Innovations." *Minerva* 56 (2018): 109–34. https://doi.org/10.1007/s11024-018-9343-7.

Wible, Brad. "Patents from Papers Both Basic and Applied." *Science* 356, no. 6333 (2017): 37–38. https://doi.org/10.1126/science.356.6333.37-f.

Williams, Roger L. *Evan Pugh's Penn State: America's Model Agricultural College.* University Park, PA: Penn State University Press, 2018.

Windolf, Paul. *Expansion and Structural Change: Higher Education in Germany, United States, and Japan, 1870–1990.* Boulder, CO: Westview Press, 1997.

Wöhlert, Romy. "Communication in International Collaborative Research Teams," *Studies in Communication and Media* 9, no. 2 (2020): 151–217. doi.org/10.5771/2192-4007-2020-2.

The World Bank. "School Enrollment, Tertiary (% Gross)." Last modified September 2021. https://data.worldbank.org/indicator/SE.TER.ENRR. (accessed May 20, 2022).

Wu, Lingfei, Aniket Kittur, Hyejin Youn, Staša Milojević, Erin Leahey, Stephen Fiore, and Yong-Yeol Ahn. "Metrics and Mechanisms: Measuring the Unmeasurable in the Science of Science." *Journal of Informetrics* 16 (2022): 101290. https://doi.org/10.1016/j.joi.2022.101290.

Wu, Lingfei, Dashun Wang, and James A. Evans. "Large Teams Develop and Small Teams Disrupt Science and Technology." *Nature* 566 (2019): 378–82. https://doi.org/10.1038/s41586-019-0941-9.

Xie, Yu, and Alexandra A. Killewald. *Is American Science in Decline?* Cambridge, MA: Harvard University Press, 2012.

Yu, Wan. "The Collaboration Network of American and Korean Universities." PhD dissertation, Pennsylvania State University, 2023. https://etda.libraries.psu.edu/files/final_submissions/28573.

Zapp, Mike. "The Legitimacy of Science and the Populist Backlash: Cross-National and Longitudinal Trends and Determinants of Attitudes Toward Science."*Public Understanding of Science* 31 (2022): 885–902. https://doi.org/10.1177/09636625221093897.

———. "Revisiting the Global Knowledge Economy: The Worldwide Expansion of Research and Development Personnel, 1980–2015." *Minerva* 60 (2022): 181–208. https://doi.org/10.1007/s11024-021-09455-4.

Zapp, Mike, and Julia C. Lerch. "Imagining the World: Conceptions and Determinants of Internationalization in Higher Education Curricula Worldwide." *Sociology of Education* 93 (2020): 372–92. https://doi.org/10.1177/0038040720929304.

Zapp, Mike, Marcelo Marques, and Justin J.W. Powell. *European Educational Research (Re)Constructed: Institutional Change in Germany, the United Kingdom, Norway, and the European Union.* Oxford, UK: Symposium Books, 2018.

Zapp, Mike, and Justin J.W. Powell. "Moving Towards Mode 2? Evidence-Based Policy-

Making and the Changing Conditions for Educational Research in Germany." *Science and Public Policy* 44, no. 5 (2017): 645–55. https://doi.org/10.1093/scipol/scw091.

Zhang, Han, Donald Patton, and Martin Kenney. "Building Global-Class Universities: Assessing the Impact of the 985 Project." *Research Policy* 42, no. 3 (2013): 765–75. https://doi.org/10.1016/j.respol.2012.10.003.

Zhang, Liang, Justin J.W. Powell, and David P. Baker. "Exponential Growth and the Shifting Global Center of Gravity of Science Production, 1900–2011." *Change: The Magazine of Higher Learning* 47, no. 4 (2015): 46–49. https://doi.org/10.1080/00091383.2015 .1053777.

Zhang, Liang, Liang Sun, and Wei Bao. "The Rise of Higher Education and Science in China." In *The Century of Science: The Global Triumph of the Research University*, ed. Justin J.W. Powell, David P. Baker, and Frank Fernandez, 141–72. Bingley, UK: Emerald, 2017. https://doi.org/10.1108/S1479-367920170000033008.

Zhang, Lin. "Understanding Chinese Science: New Perspectives from Scientometrics and Research Policy." Lecture presented in the University of Luxembourg Science of Science Lecture Series, virtual, July 21, 2021. https://wwwen.uni.lu/research/fhse/dsoc/news_ events/science_of_science_in_the_spotlight.

Zippel, Kathrin. *Women in Global Science: Advancing Academic Careers Through International Collaboration*. Stanford, CA: Stanford University Press, 2017.

"Zombie Research Haunts Academic Literature Long After Its Supposed Demise." *The Economist,* June 26, 2021. https://www.economist.com/graphic-detail/2021/06/26/ zombie-research-haunts-academic-literature-long-after-its-supposed-demise.

Index

Printed in the USA
CPSIA information can be obtained
at www.ICGtesting.com
JSHW021028040324
58400JS00013B/4